Yamaha Vision Owners Workshop Manual

by Curt Choate
and John H Haynes Member of the Guild of Motoring Writers

Models covered
Yamaha XZ 550 Vision
Introduced UK and USA 1982

ISBN 978 1 85010 761 3

© **Haynes Publications, Inc. 1991**
With permission from J.H. Haynes & Co., Ltd.
All rights reserved. No part of this book may be reproduced or transmitted in
any form or by any means, electronic or mechanical, including photocopying,
recording or by any information storage or retrieval system, without permission
in writing from the copyright holder

(821-3T3)

J H Haynes & Co. Ltd.
Haynes North America, Inc

www.haynes.com

Acknowledgements

Our thanks to Mitsui Machinery Sales (UK) Ltd for permission to reproduce certain illustrations used in this manual. We would also like to thank NGK Spark Plugs (UK) Ltd for supplying the color spark plug condition photos.

About this manual

Its purpose

The purpose of this manual is to help you maintain and repair your motorcycle. It can do so in several ways. It can help you decide what work must be done, even if you choose to have it done by a dealer service department or a repair shop, it provides information and procedures for routine maintenance and it offers diagnostic and repair procedures to follow when trouble occurs.

It is hoped that you will use the manual to tackle the work yourself. For many simple jobs, doing it yourself may be quicker than arranging an appointment to get the machine into a shop and making the trips to leave it and pick it up. More importantly, a lot of money can be saved by avoiding the expense the shop must pass on to you to cover its labor and overhead costs. An added benefit is the sense of satisfaction and accomplishment that you feel after having done the job yourself.

Using the manual

The manual is divided into Chapters. Each Chapter is divided into numbered Sections which are headed in bold type between horizontal lines. Each Section consists of consecutively numbered paragraphs.

The two types of illustrations used (figures and photographs) are referenced by a number preceding their caption. Figure reference numbers denote Chapter and numerical sequence within the Chapter (i.e. Fig. 3.4 means Chapter 3, figure number 4). Figure captions are followed by a Section number which ties the figure to a specific portion of the text. All photographs apply to the Chapter in which they appear and the reference number pinpoints the pertinent Section and paragraph.

Procedures, once described in the text, are not normally repeated. When it is necessary to refer to another Chapter, the reference will be given as Chapter and Section number (i.e. Chapter 1, Section 16). Cross references given without use of the word ''Chapter'' apply to Sections and/or paragraphs in the same Chapter. For example, ''see Section 8'' means in the same Chapter.

Reference to the left or right side of the motorcycle is based on the assumption that one is sitting on the seat, facing forward.

Motorcycle manufacturers continually make changes to specifications and recommendations, and these, when notified, are incorporated into our manuals at the earliest opportunity.

Even though extreme care has been taken during the preparation of this manual, neither the publisher nor the author can accept responsibility for any errors in, or omissions from, the information given.

Introduction to the Yamaha Vision

The Yamaha XZ550 Vision is an innovative motorcycle, unique in its displacement category. The engine is a 70-degree V-twin that is liquid cooled and equipped with a counter-balancer system designed to cancel the vibrations inherent in this type of engine. Each cylinder head has dual overhead camshafts and four valves (two intake and two exhaust). A five-speed, constant mesh transmission and shaft final drive transmit power to the rear wheel.

The patented YICS (Yamaha Induction Control System), designed to improve combustion and increase fuel economy, is standard equipment, along with downdraft, automotive-style carburetors.

The "hang-support" frame allows for lower than normal engine placement and center of gravity, while maintaining the rigidity necessary for precise handling. The trailing axle front forks help to maintain a short wheelbase, which gives the machine its quick handling characteristics. The rear suspension is Yamaha's proven Monoshock, which has five-way adjustable spring preload.

The Vision was introduced in the USA as a 1982 (RJ) model. For the 1983 model year (its last), a fairing was added, several minor changes/improvements were made and the designation was changed to RK.

Contents

1982 Yamaha XZ 550 RJ Vision

Identification numbers

Yamaha motorcycles have a Vehicle Identification Number (VIN) label attached to the left side or front side of the steering head. This number is usually required for registration and licensing of the vehicle.

The frame serial number is stamped into the right side of the steering head and the engine serial number is stamped into the right engine case. Both of these numbers should be recorded and kept in a safe place so they can be furnished to law enforcement officials in the event of theft.

The frame serial number, engine serial number and carburetor identification number should also be kept in a handy place (such as with your driver's license) so they are always available when purchasing or ordering parts for your machine.

Frame serial number location

Engine serial number location

Buying parts

Once you have found all the identification numbers, record them for reference when buying parts. Since the manufacturers change specifications, parts and vendors (companies that manufacture various components on the machine), providing the ID numbers is the only way to be reasonably sure that you are buying the correct parts.

Whenever possible, take the worn part to the dealer so direct comparison with the new component can be made. Along the trail from the manufacturer to the parts shelf, there are numerous places that the part can end up with the wrong number or be listed incorrectly.

The two places to purchase new parts for your motorcycle—the accessory store and the franchised dealer—differ in the type of parts that they carry. While dealers can obtain virtually every part for your cycle, the accessory dealer is usually limited to normal high wear items such as shock absorbers, tune-up parts, various engine gaskets, cables, brake parts, etc. Rarely will an accessory outlet have major suspension components, cylinders, transmission gears, or cases.

Used parts can be obtained for roughly half the price of new ones, but you can't always be sure of what you're getting. Once again, take your worn part to the wrecking yard for direct comparison.

Whether buying new, used or rebuilt parts, the best course is to deal directly with someone who specializes in parts for your particular make.

General specifications

Frame and suspension

Wheelbase	56.9 in (1445 mm)
Overall length	83.5 in (2120 mm)
Overall width	33.3 in (845 mm)
Overall height	43.7 in (1110 mm)
Seat height	30.7 in (780 mm)
Weight (with oil and full fuel tank)	467 lbs (212 Kg)
Front suspension	Telescopic trailing axle fork
Rear suspension	Monoshock
Front brake	Hydraulic disc
Rear brake	Drum
Fuel capacity	4.5 US gal (17 liters)

Engine

Type	Liquid cooled, 4-stroke, DOHC V-twin
Displacement	552 cc
Ignition system	TCI (Transistor Controlled Ignition)
Carburetor type	Downdraft automotive type
Clutch	Wet, multi-plate
Transmission	5-speed, constant mesh

Maintenance techniques,
tools and working facilities

Basic maintenance techniques

There are a number of techniques involved in maintenance and repair that will be referred to throughout this manual. Application of these techniques will enable the amateur mechanic to be more efficient, better organized and capable of performing the various tasks properly, which will ensure that the repair job is thorough and complete.

Fastening systems

Fasteners, basically, are nuts, bolts and screws used to hold two or more parts together. There are a few things to keep in mind when working with fasteners. Almost all of them use a locking device of some type (either a lock washer, locknut, locking tab or thread adhesive). All threaded fasteners should be clean, straight, have undamaged threads and undamaged corners on the hex head where the wrench fits. Develop the habit of replacing all damaged nuts and bolts with new ones.

Rusted nuts and bolts should be treated with a penetrating fluid to ease removal and prevent breakage. Some mechanics use turpentine in a spout type oil can, which works quite well. After applying the rust penetrant, let it ''work'' for a few minutes before trying to loosen the nut or bolt. Badly rusted fasteners may have to be chiseled off or removed with a special nut breaker, available at tool stores.

Flat washers and lock washers, when removed from an assembly, should always be replaced exactly as removed. Replace any damaged washers with new ones. Always use a flat washer between a lock washer and any soft metal surface (such as aluminum), thin sheet metal or plastic. Special locknuts can only be used once or twice before they lose their locking ability and must be replaced.

If a bolt or stud breaks off in an assembly, it can be drilled out and removed with a special tool called an E-Z out. Most dealer service departments and motorcycle repair shops can perform this task, as well as others (such as the repair of threaded holes that have been stripped out).

Torquing sequences and procedures

When threaded fasteners are tightened, they are often tightened to a specific torque value (torque is basically a twisting force). Over tightening the fastener can weaken it and cause it to break, while under tightening can cause it to eventually come loose. Each bolt, depending on the material it's made of, the diameter of its shank and the material it is threaded into, has a specific torque value, which is noted in the Specifications. Be sure to follow the torque recommendations closely. For fasteners not requiring a specific torque, a general torque value chart is presented as a guide.

Fasteners laid out in a pattern (i.e. cylinder head bolts, engine case bolts, etc.) must be loosened or tightened in a sequence to avoid warping the component. Initially, the bolts/nuts should go on finger tight only. Next, they should be tightened one full turn each, in a criss-cross or diagonal pattern. After each one has been tightened one full turn, return to the first one tightened and tighten them all one half turn, following the same pattern. Finally, tighten each of them one quarter turn at a time until each fastener has been tightened to the proper torque. To loosen and remove the fasteners the procedure would be reversed.

Disassembly sequence

Component disassembly should be done with care and purpose to help ensure that the parts go back together properly during reassembly. Always keep track of the sequence in which parts are removed. Take note of special characteristics or marks on parts that can be installed more than one way (such as a grooved thrust washer on a shaft). It's a good idea to lay the disassembled parts out on a clean surface in the order that they were removed. It may also be helpful to make sketches or take instant photos of components before removal.

When removing fasteners from a component, keep track of their locations. Sometimes threading a bolt back in a part, or putting the washers and nut back on a stud, can prevent mixups later. If nuts and bolts cannot be returned to their original locations, they should be kept in a compartmented box or a series of small boxes. A cupcake or muffin tin is ideal for this purpose, since each cavity can hold the bolts and nuts from a particular area (i.e. engine case bolts, valve cover bolts, engine mount bolts, etc.). A pan of this type is especially helpful when working on assemblies with very small parts (such as the carburetors and the valve train). The cavities can be marked with paint or tape to identify the contents.

Whenever wiring looms, harnesses or connectors are separated, it's a good idea to identify the two halves with numbered pieces of masking tape so they can be easily reconnected.

Gasket sealing surfaces

Throughout any motorcycle, gaskets are used to seal the mating surfaces between components and keep lubricants, fluids, vacuum or pressure contained in an assembly.

Many times these gaskets are coated with a liquid or paste type gasket sealing compound before assembly. Age, heat and pressure can sometimes cause the two parts to stick together so tightly that they are very difficult to separate. In most cases, the part can be loosened by striking it with a soft-faced hammer near the mating surfaces. A regular hammer can be used if a block of wood is placed between the hammer and the part. Do not hammer on cast parts or parts that could be easily damaged. With any particularly stubborn part, always recheck to make sure that every fastener has been removed.

Avoid using a screwdriver or bar to pry apart components, as they can easily mar the gasket sealing surfaces of the parts (which must remain smooth). If prying is absolutely necessary, use a piece of wood, but keep in mind that extra clean-up will be necessary if the wood splinters.

After the parts are separated, the old gasket must be carefully scraped off and the gasket surfaces cleaned. Stubborn gasket material can be soaked with a gasket remover (available in aerosol cans) to soften it so it can be easily scraped off. A scraper can be fashioned from a piece of copper tubing by flattening and sharpening one end. Copper is recommended because it is usually softer than the surfaces to be scraped, which reduces the chance of gouging the part. Some gaskets can be removed with a wire brush, but regardless of the method used, the mating surfaces must be left clean and smooth. If for some reason the gasket surface is gouged, then a gasket sealer thick enough to fill scratches will have to be used during reassembly of the components. For most applications, a non-drying (or semi-drying) gasket sealer is best.

Hose removal tips

Hose removal precautions closely parallel gasket removal precautions. Avoid scratching or gouging the surface that the hose mates against or the connection may leak. Because of various chemical reactions, the rubber in hoses can bond itself to the metal spigot that the hose fits over. To remove a hose, first loosen the hose clamps that secure it to the spigot. Then, with slip joint pliers, grab the hose at the clamp and rotate it around the spigot. Work it back and forth until it is completely free, then pull it off (silicone or other lubricants will ease removal if they can be applied between the hose and the outside of the spigot). Apply the same lubricant to the inside of the hose and the outside of the spigot to simplify installation.

If a hose clamp is broken or damaged, do not reuse it. Also, do not reuse hoses that are cracked, split or torn.

Tools

A selection of good tools is a basic requirement for anyone who plans to maintain and repair a motorcycle. For the owner who has few tools, if any, the initial investment might seem high, but when compared to the spiraling costs of routine maintenance and repair, it is a wise one.

To help the owner decide which tools are needed to perform the tasks detailed in this manual, the following tool lists are offered: *Maintenance and minor repair, Repair and overhaul* and *Special*. The newcomer to practical mechanics should start off with the *Maintenance and minor repair* tool kit, which is adequate for the simpler jobs. Then, as confidence and experience grow, the owner can tackle more difficult tasks, buying additional tools as they are needed. Eventually the basic kit will be built into the *Repair and overhaul* tool set. Over a period of time, the experienced do-it-yourselfer will assemble a tool set complete enough for most repair and overhaul procedures and will add tools from the *Special* category when it is felt that the expense is justified by the frequency of use.

Maintenance and minor repair tool kit

The tools in this list should be considered the minimum required for performance of routine maintenance, servicing and minor repair work. We recommend the purchase of combination wrenches (box end and open end combined in one wrench); while more expensive than open-ended ones, they offer the advantages of both types of wrench.

> *Combination wrench set (6 mm to 22 mm)*
> *Adjustable wrench — 8 in*
> *Spark plug socket (with rubber insert)*
> *Spark plug gap adjusting tool*
> *Feeler gauge set*
> *Standard screwdriver (5/16 in x 6 in)*
> *Phillips screwdriver (No. 2 x 6 in)*
> *Combination (slip-joint) pliers — 6 in*
> *Hacksaw and assortment of blades*
> *Tire pressure gauge*
> *Control cable pressure luber*
> *Grease gun*
> *Oil can*
> *Fine emery cloth*
> *Wire brush*
> *Hand impact screwdriver and bits*
> *Funnel (medium size)*
> *Safety goggles*
> *Drain pan*

Note: *Since basic ignition timing checks are a part of routine maintenance, it will be necessary to purchase a good quality, inductive pick-up stroboscopic timing light. Although it is included in the list of Special tools, it is mentioned here because ignition timing checks cannot be made without one.*

Repair and overhaul tool set

These tools are essential for anyone who plans to perform major repairs and are intended to supplement those in the *Maintenance and minor repair* tool kit. Included is a comprehensive set of sockets which, though expensive, are invaluable because of their versatility (especially when various extensions and drives are available). We recommend the 3/8 inch drive over the 1/2 inch drive for general motorcycle maintenance and repair (ideally, the mechanic would have a 3/8 inch drive set and a 1/2 inch drive set).

> *Socket set(s)*
> *Reversible ratchet*
> *Extension — 6 in*
> *Universal joint*
> *Torque wrench (same size drive as sockets)*
> *Ball pein hammer — 8 oz*
> *Soft-faced hammer (plastic/rubber)*
> *Standard screwdriver (1/4 in x 6 in)*
> *Standard screwdriver (stubby — 5/16 in)*

> *Phillips screwdriver (No. 3 x 8 in)*
> *Phillips screwdriver (stubby — No. 2)*
> *Pliers — vise grip*
> *Pliers — lineman's*
> *Pliers — needle nose*
> *Pliers — snap-ring (internal and external)*
> *Cold chisel — 1/2 in*
> *Scriber*
> *Scraper (made from flattened copper tubing)*
> *Center punch*
> *Pin punches (1/16, 1/8, 3/16 in)*
> *Steel rule/straightedge — 12 in*
> *Allen wrench set (4 mm to 10 mm)*
> *Pin-type spanner wrench*
> *A selection of files*
> *Wire brush (large)*

Note: *Another tool which is often useful is an electric drill motor with a chuck capacity of 3/8 inch (and a set of good quality drill bits).*

Special tools

The tools in this list include those which are not used regularly, are expensive to buy, or which need to be used in accordance with their manufacturer's instructions. Unless these tools will be used frequently, it is not very economical to purchase many of them. A consideration would be to split the cost and use between yourself and a friend or friends (i.e. members of a motorcycle club).

This list primarily contains tools and instruments widely available to the public, as well as some special tools produced by the vehicle manufacturer for distribution to dealer service departments. As a result, references to the manufacturer's special tools are occasionally included in the text of this manual. Generally, an alternative method of doing the job without the special tool is offered. However, sometimes there is no alternative to their use. Where this is the case, and the tool cannot be purchased or borrowed, the work should be turned over to the dealer service department or a motorcycle repair shop.

> *Valve spring compressor*
> *Piston ring removal and installation tool*
> *Piston pin puller*
> *Telescoping gauges*
> *Micrometer(s) and/or dial/Vernier calipers*
> *Cylinder surfacing hone*
> *Cylinder compression gauge*
> *Dial indicator set*
> *Multimeter*
> *Valve lifter depressing tool*
> *TORX driver bits (T-25 and T-30)*
> *Flywheel (alternator rotor) puller*
> *46 mm (1-13/16 inch) crowfoot wrench or deep socket*
> *Manometer or vacuum gauge set*
> *Stroboscopic timing light*
> *Work light with extension cord*
> *Small air compressor with blow gun and tire chuck*

Buying tools

For the do-it-yourselfer who is just starting to get involved in motorcycle maintenance and repair, there are a number of options available when purchasing tools. If maintenance and minor repair is the extent of the work to be done, the purchase of individual tools is satisfactory. If, on the other hand, extensive work is planned, it would be a good idea to purchase a modest tool set from one of the large retail chain stores. A set can usually be bought at a substantial savings over the individual tool prices (and they often come with a tool box). As additional tools are needed, add-on sets, individual tools and a larger tool box can be purchased to expand the tool selection. Building a tool set gradually allows the cost of the tools to be spread over a longer period of time and gives the mechanic the freedom to choose only those tools that will actually be used.

Tool stores and motorcycle dealers will often be the only source of some of the special tools that are needed, but regardless of where tools are bought, try to avoid cheap ones (especially when buying screwdrivers and sockets) because they won't last very long. The expense involved in replacing cheap tools will eventually be greater than the initial cost of quality tools.

Spark plug gap adjusting tool

Feeler gauge set

Control cable pressure luber

Hand impact screwdriver and bits

Care and maintenance of tools

Good tools are expensive, so it makes sense to treat them with respect. Keep them clean and in usable condition and store them properly when not in use. Always wipe off any dirt, grease or metal chips before putting them away. Never leave tools lying around in the work area.

Some tools, such as screwdrivers, pliers, wrenches and sockets, can be hung on a panel mounted on the garage or workshop wall, while others should be kept in a tool box or tray. Measuring instruments, gauges, meters, etc. must be carefully stored where they cannot be damaged by weather or impact from other tools.

When tools are used with care and stored properly, they will last a very long time. Even with the best of care, tools will wear out if used frequently. When a tool is damaged or worn out, replace it; subsequent jobs will be safer and more enjoyable if you do.

Working facilities

Not to be overlooked when discussing tools is the workshop. If anything more than routine maintenance is to be carried out, some sort of suitable work area is essential.

It is understood, and appreciated, that many home mechanics do not have a good workshop or garage available and end up removing an engine or doing major repairs outside (it is recommended, however, that the overhaul or repair be completed under the cover of a roof).

A clean, flat workbench or table of comfortable working height is an absolute necessity. The workbench should be equipped with a vise that has a jaw opening of at least four inches.

As mentioned previously, some clean, dry storage space is also required for tools, as well as the lubricants, fluids, cleaning solvents, etc. which soon become necessary.

Sometimes waste oil and fluids, drained from the engine or cooling system during normal maintenance or repairs, present a disposal problem. To avoid pouring them on the ground or into a sewage system, simply pour the used fluids into large containers, seal them with caps and take them to an authorized disposal site or service station. Plastic jugs (such as old antifreeze containers) are ideal for this purpose.

Always keep a supply of old newspapers and clean rags available. Old towels are excellent for mopping up spills. Many mechanics use rolls of paper towels for most work because they are readily available and disposable. To help keep the area under the motorcycle clean, a large cardboard box can be cut open and flattened to protect the garage or shop floor.

Whenever working over a painted surface (such as the fuel tank) cover it with an old blanket or bedspread to protect the finish.

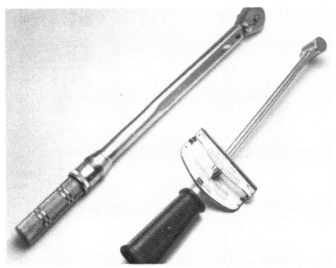

Torque wrenches (left—click type; right—beam type)

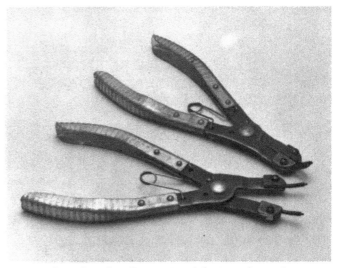

Snap-ring pliers (top—external; bottom—internal)

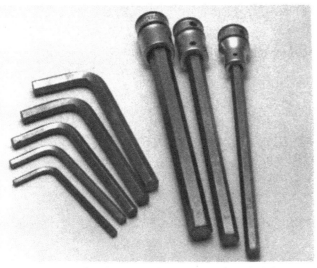

Allen wrenches (left) and Allen head sockets (right)

Valve spring compressor

Piston ring removal/installation tool

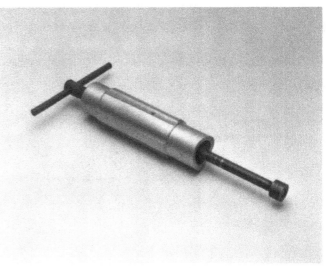

Piston pin puller

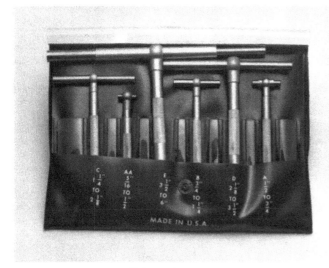

Telescoping gauges

0-to-1 inch micrometer

Cylinder surfacing hone

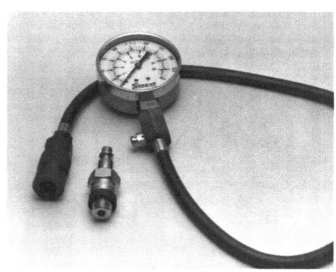

Cylinder compression gauge

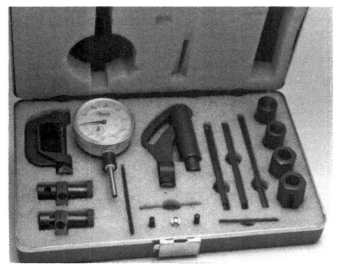

Dial indicator set

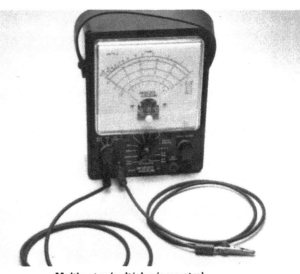

Multimeter (volt/ohm/ammeter)

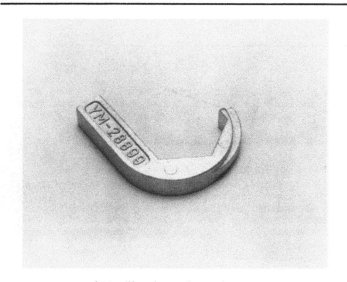

Valve lifter depressing tool

TORX driver bit (T-25)

Alternator rotor (flywheel) puller

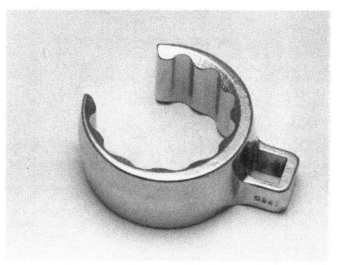

46 mm (1-13/16 inch) crowfoot wrench

Safety first!

Professional motor mechanics are trained in safe working procedures. However enthusiastic you may be about getting on with the job in hand, do take the time to ensure that your safety is not put at risk. A moment's lack of attention can result in an accident, as can failure to observe certain elementary precautions.

There will always be new ways of having accidents, and the following points do not pretend to be a comprehensive list of all dangers; they are intended rather to make you aware of the risks and to encourage a safety-conscious approach to all work you carry out on your vehicle.

Essential DOs and DON'Ts

DON'T start the engine without first ascertaining that the transmission is in neutral.

DON'T suddenly remove the filler cap from a hot cooling system – cover it with a cloth and release the pressure gradually first, or you may get scalded by escaping coolant.

DON'T attempt to drain oil until you are sure it has cooled sufficiently to avoid scalding you.

DON'T grasp any part of the engine, exhaust or silencer without first ascertaining that it is sufficiently cool to avoid burning you.

DON'T allow brake fluid or antifreeze to contact the machine's paintwork or plastic components.

DON'T syphon toxic liquids such as fuel, brake fluid or antifreeze by mouth, or allow them to remain on your skin.

DON'T inhale dust – it may be injurious to health (see *Asbestos* heading).

DON'T allow any spilt oil or grease to remain on the floor – wipe it up straight away, before someone slips on it.

DON'T use ill-fitting spanners or other tools which may slip and cause injury.

DON'T attempt to lift a heavy component which may be beyond your capability – get assistance.

DON'T rush to finish a job, or take unverified short cuts.

DON'T allow children or animals in or around an unattended vehicle.

DON'T inflate a tyre to a pressure above the recommended maximum. Apart from overstressing the carcase and wheel rim, in extreme cases the tyre may blow off forcibly.

DO ensure that the machine is supported securely at all times. This is especially important when the machine is blocked up to aid wheel or fork removal.

DO take care when attempting to slacken a stubborn nut or bolt. It is generally better to pull on a spanner, rather than push, so that if slippage occurs you fall away from the machine rather than on to it.

DO wear eye protection when using power tools such as drill, sander, bench grinder etc.

DO use a barrier cream on your hands prior to undertaking dirty jobs – it will protect your skin from infection as well as making the dirt easier to remove afterwards; but make sure your hands aren't left slippery. Note that long-term contact with used engine oil can be a health hazard.

DO keep loose clothing (cuffs, tie etc) and long hair well out of the way of moving mechanical parts.

DO remove rings, wristwatch etc, before working on the vehicle – especially the electrical system.

DO keep your work area tidy – it is only too easy to fall over articles left lying around.

DO exercise caution when compressing springs for removal or installation. Ensure that the tension is applied and released in a controlled manner, using suitable tools which preclude the possibility of the spring escaping violently.

DO ensure that any lifting tackle used has a safe working load rating adequate for the job.

DO get someone to check periodically that all is well, when working alone on the vehicle.

DO carry out work in a logical sequence and check that everything is correctly assembled and tightened afterwards.

DO remember that your vehicle's safety affects that of yourself and others. If in doubt on any point, get specialist advice.

IF, in spite of following these precautions, you are unfortunate enough to injure yourself, seek medical attention as soon as possible.

Asbestos

Certain friction, insulating, sealing, and other products – such as brake linings, clutch linings, gaskets, etc – contain asbestos. *Extreme care must be taken to avoid inhalation of dust from such products since it is hazardous to health.* If in doubt, assume that they *do* contain asbestos.

Fire

Remember at all times that petrol (gasoline) is highly flammable. Never smoke, or have any kind of naked flame around, when working on the vehicle. But the risk does not end there – a spark caused by an electrical short-circuit, by two metal surfaces contacting each other, by careless use of tools, or even by static electricity built up in your body under certain conditions, can ignite petrol vapour, which in a confined space is highly explosive.

Always disconnect the battery earth (ground) terminal before working on any part of the fuel or electrical system, and never risk spilling fuel on to a hot engine or exhaust.

It is recommended that a fire extinguisher of a type suitable for fuel and electrical fires is kept handy in the garage or workplace at all times. Never try to extinguish a fuel or electrical fire with water.

Note: *Any reference to a 'torch' appearing in this manual should always be taken to mean a hand-held battery-operated electric lamp or flashlight. It does **not** mean a welding/gas torch or blowlamp.*

Fumes

Certain fumes are highly toxic and can quickly cause unconsciousness and even death if inhaled to any extent. Petrol (gasoline) vapour comes into this category, as do the vapours from certain solvents such as trichloroethylene. Any draining or pouring of such volatile fluids should be done in a well ventilated area.

When using cleaning fluids and solvents, read the instructions carefully. Never use materials from unmarked containers – they may give off poisonous vapours.

Never run the engine of a motor vehicle in an enclosed space such as a garage. Exhaust fumes contain carbon monoxide which is extremely poisonous; if you need to run the engine, always do so in the open air or at least have the rear of the vehicle outside the workplace.

The battery

Never cause a spark, or allow a naked light, near the vehicle's battery. It will normally be giving off a certain amount of hydrogen gas, which is highly explosive.

Always disconnect the battery earth (ground) terminal before working on the fuel or electrical systems.

If possible, loosen the filler plugs or cover when charging the battery from an external source. Do not charge at an excessive rate or the battery may burst.

Take care when topping up and when carrying the battery. The acid electrolyte, even when diluted, is very corrosive and should not be allowed to contact the eyes or skin.

If you ever need to prepare electrolyte yourself, always add the acid slowly to the water, and never the other way round. Protect against splashes by wearing rubber gloves and goggles.

Mains electricity and electrical equipment

When using an electric power tool, inspection light etc, always ensure that the appliance is correctly connected to its plug and that, where necessary, it is properly earthed (grounded). Do not use such appliances in damp conditions and, again, beware of creating a spark or applying excessive heat in the vicinity of fuel or fuel vapour. Also ensure that the appliances meet the relevant national safety standards.

Ignition HT voltage

A severe electric shock can result from touching certain parts of the ignition system, such as the HT leads, when the engine is running or being cranked, particularly if components are damp or the insulation is defective. Where an electronic ignition system is fitted, the HT voltage is much higher and could prove fatal.

Motorcycle chemicals and lubricants

A number of chemicals and lubricants are available for use in motorcycle maintenance and repair. They include a wide variety of products ranging from cleaning solvents and degreasers to lubricants and protective sprays for rubber, plastic and vinyl.

Contact point/spark plug cleaner is a solvent used to clean oily film and dirt from points, grime from electrical connectors and oil deposits from spark plugs. It is oil free and leaves no residue. It can also be used to remove gum and varnish from carburetor jets and other orifices.

Carburetor cleaner is similar to contact point/spark plug cleaner but it usually has a stronger solvent and may leave a slight oily reside. It is not recommended for cleaning electrical components or connections.

Brake system cleaner is used to remove grease or brake fluid from brake system components (where clean surfaces are absolutely necessary and petroleum-based solvents cannot be used); it also leaves no residue.

Silicone-based lubricants are used to protect rubber parts such as hoses and grommets, and are used as lubricants for hinges and locks.

Multi-purpose grease is an all purpose lubricant used wherever grease is more practical than a liquid lubricant such as oil. Some multi-purpose grease is colored white and specially formulated to be more resistant to water than ordinary grease.

Gear oil (sometimes called gear lube) is a specially designed oil used in transmissions and final drive units, as well as other areas where high friction, high temperature lubrication is required. It is available in a number of viscosities (weights) for various applications.

Motor oil, of course, is the lubricant specially formulated for use in the engine. It normally contains a wide variety of additives to prevent corrosion and reduce foaming and wear. Motor oil comes in various weights (viscosity ratings) of from 5 to 80. The recommended weight of the oil depends on the seasonal temperature and the demands on the engine. Light oil is used in cold climates and under light load conditions; heavy oil is used in hot climates and where high loads are encountered. Multi-viscosity oils are designed to have characteristics of both light and heavy oils and are available in a number of weights from 5W-20 to 20W-50.

Gas additives perform several functions, depending on their chemical makeup. They usually contain solvents that help dissolve gum and varnish that build up on carburetor and intake parts. They also serve to break down carbon deposits that form on the inside surfaces of the combustion chambers. Some additives contain upper cylinder lubricants for valves and piston rings.

Brake fluid is a specially formulated hydraulic fluid that can withstand the heat and pressure encountered in brake systems. Care must be taken that this fluid does not come in contact with painted surfaces or plastics. An opened container should always be resealed to prevent contamination by water or dirt.

Chain lubricants are formulated especially for use on motorcycle final drive chains. A good chain lube should adhere well and have good penetrating qualities to be effective as a lubricant inside the chain and on the side plates, pins and rollers. Most chain lubes are either the foaming type or quick drying type and are usually marketed as sprays.

Degreasers are heavy duty solvents used to remove grease and grime that may accumulate on engine and frame components. They can be sprayed or brushed on and, depending on the type, are rinsed with either water or solvent.

Solvents are used alone or in combination with degreasers to clean parts and assemblies during repair and overhaul. The home mechanic should use only solvents that are non-flammable and that do not produce irritating fumes.

Gasket sealing compounds may be used in conjunction with gaskets, to improve their sealing capabilities, or alone, to seal metal-to-metal joints. Many gasket sealers can withstand extreme heat, some are impervious to gasoline and lubricants, while others are capable of filling and sealing large cavities. Depending on the intended use, gasket sealers either dry hard or stay relatively soft and pliable. They are usually applied by hand, with a brush, or are sprayed on the gasket sealing surfaces.

Thread cement is an adhesive locking compound that prevents threaded fasteners from loosening because of vibration. It is available in a variety of types for different applications.

Moisture dispersants are usually sprays that can be used to dry out electrical components such as the fuse block and wiring connectors. Some types can also be used as treatment for rubber and as a lubricant for hinges, cables and locks.

Waxes and polishes are used to help protect painted and plated surfaces from the weather. Different types of paint may require the use of different types of wax polish. Some polishes utilize a chemical or abrasive cleaner to help remove the top layer of oxidized (dull) paint on older vehicles. In recent years, many non-wax polishes (that contain a wide variety of chemicals such as polymers and silicones) have been introduced. These non-wax polishes are usually easier to apply and last longer than conventional waxes and polishes.

Troubleshooting

Contents

Engine doesn't start or is difficult to start

1 Starter motor does not rotate

1 Engine stop switch Off.
2 Fuse blown. Check fuse block under the seat (Chapter 8).
3 Battery voltage low. Check and recharge battery (Chapter 8).
4 Starter motor defective. Make sure that the wiring to the starter is secure. Make sure the starter solenoid (relay) clicks when the start button is pushed. If the solenoid clicks, then the fault is in the wiring or motor.
5 Starter solenoid (relay) faulty. It is located behind the left side cover. Check it according to the procedure in Chapter 8.
6 Starter button not contacting. The contacts could be wet, corroded or dirty. Disassemble and clean the switch (Chapter 8).
7 Wiring open or shorted. Check all wiring connections and wiring looms to make sure that they are dry, tight and not corroded. Also check for broken or frayed wires that can cause a short to ground (see wiring diagram, Chapter 8).
8 Ignition switch defective. Check the switch according to the pro-

cedure in Chapter 8. Replace the switch with a new one if it is defective.
9 Engine stop switch defective. Check for wet, dirty or corroded contacts. Clean or replace the switch as necessary (Chapter 8).
10 Faulty neutral switch. Check the wiring to the switch and the switch itself according to the procedures in Chapter 8.

2 Starter motor rotates but engine does not turn over

1 Starter motor clutch defective. Inspect and repair or replace (Chapter 2).
2 Damaged idler or starter gears. Inspect and replace the damaged parts (Chapter 2).

3 Starter works but engine won't turn over (seized)

Seized engine caused by one or more internally damaged components. Failure due to wear, abuse or lack of lubrication. Damage can include seized valves, valve lifters, camshaft, pistons, crankshaft, connecting rod bearings, or transmission gears or bearings. Refer to Chapter 2 for engine disassembly.

4 No fuel flow

1 No fuel in tank.
2 Fuel petcock vacuum hose broken or disconnected.
3 Tank cap air vent obstructed. Usually caused by dirt or water. Remove it and clean the cap vent hole.
4 Fuel petcock clogged. Remove the petcock and clean it and the filter (Chapter 1).
5 Fuel line clogged. Pull the fuel line loose and carefully blow through it.
6 Inlet needle valves clogged. For both of the valves to be clogged, either a very bad batch of fuel with an unusual additive has been used, or some other foreign object has entered the tank. Many times after a machine has been stored for many months without running, the fuel turns to a varnish-like liquid and forms deposits on the inlet needle valves and jets. The carburetor should be removed and overhauled if draining the float bowls does not alleviate the problem.

5 Engine flooded

1 Float level too high. Check and adjust as described in Chapter 4.
2 Inlet needle valve worn or stuck open. A piece of dirt, rust or other debris can cause the inlet needle to seat improperly, causing excess fuel to be admitted to the float bowl. In this case, the float chamber should be cleaned and the needle and seat inspected. If the needle and seat are worn, then the leaking will persist and the parts should be replaced with new ones (Chapter 4).
3 Starting technique faulty. Under normal circumstances (i.e., if all the carburetor functions are sound) the machine should start with little or no throttle. When the engine is cold, the choke should be operated and the engine started without opening the throttle. When the engine is at operating temperature, only a very slight amount of throttle should be necessary. On this model, an accelerator pump is fitted and repeated opening and closing of the throttle with the engine off, will cause flooding. If the engine is flooded, turn the fuel petcock off and hold the throttle open while cranking the engine. This will allow additional air to reach the cylinders. Remember to turn the gas back on after the engine starts.

6 No spark or weak spark

1 Ignition switch Off.
2 Engine stop switch turned to the Off position.
3 Battery voltage low. Check and recharge battery as necessary (Chapter 8).
4 Spark plug dirty, defective or worn out. Locate reason for fouled plug(s) using spark plug condition chart and follow the plug maintenance procedures in Chapter 1.
5 Spark plug cap or high-tension wiring faulty. Check condition. Replace either or both components if cracks or deterioration are evident (Chapter 5).
6 Spark plug cap not making good contact. Make sure that the plug cap fits snugly over the plug end.
7 TCI unit defective. Check the unit, referring to Chapter 5 for details.
8 Pickup coil defective. Check the unit, referring to Chapter 5 for details.
9 Ignition coil defective. Check the coils, referring to Chapter 5.
10 Ignition or stop switch shorted. This is usually caused by water, corrosion, damage or excessive wear. The switches can be disassembled and cleaned with electrical contact cleaner. If cleaning does not help, replace the switches (Chapter 8).
11 Wiring shorted or broken between:
   a) Ignition switch and engine stop switch
   b) TCI unit and engine stop switch
   c) TCI unit and ignition coil
   d) Ignition coil and plug
   e) TCI unit and pickup coils
Make sure that all wiring connections are clean, dry and tight. Look for chafed and broken wires (Chapters 5 and 8).

7 Compression low

1 Spark plug loose. Remove the plug and inspect the threads. Reinstall and tighten to the specified torque (Chapter 1).
2 Cylinder head not sufficiently tightened down. If the cylinder head is suspected of being loose, then there's a chance that the gasket or head is damaged if the problem has persisted for any length of time. The head bolts should be tightened to the proper torque in the correct sequence (Chapter 2).
3 Improper valve clearance. This means that the valve is not closing completely and compression pressure is leaking past the valve. Check and adjust the valve clearances (Chapter 1).
4 Cylinder and/or piston worn. Excessive wear will cause compression pressure to leak past the rings. This is usually accompanied by worn rings as well. A top end overhaul is necessary (Chapter 2).
5 Piston rings worn, weak, broken, or sticking. Broken or sticking piston rings usually indicate a lubrication or carburetion problem that causes excess carbon deposits or seizures to form on the pistons and rings. Top end overhaul is necessary (Chapter 2).
6 Piston ring-to-groove clearance excessive. This is caused by excessive wear of the piston ring lands. Piston replacement is necessary (Chapter 2).
7 Cylinder head gasket damaged. If the heads are allowed to become loose, or if excessive carbon build-up on the piston crown and combustion chamber causes extremely high compression, the head gasket may leak. Retorquing the head is not always sufficient to restore the seal, so gasket replacement is necessary (Chapter 2).
8 Cylinder head warped. This is caused by overheating or improperly tightened head bolts. Machine shop resurfacing or head replacement is necessary (Chapter 2).
9 Valve spring broken or weak. Caused by component failure or wear; the spring(s) must be replaced (Chapter 2).
10 Valve not seating properly. This is caused by a bent valve (from over revving or improper valve adjustment), burned valve or seat (improper carburetion) or an accumulation of carbon deposits on the seat (from carburetion, lubrication problems). The valves must be cleaned and/or replaced and the seats serviced if possible (Chapter 2).

8 Stalls after starting

1 Improper choke action. Make sure the choke rod is getting a full stroke and staying in the "out" position. Adjustment of the cable slack is covered in Chapter 1.
2 Ignition malfunction. See Chapter 5.
3 Carburetor malfunction. See Chapter 4.
4 Fuel contaminated. The fuel can be contaminated with either dirt or water, or can change chemically if the machine is allowed to sit for several months or more. Drain the tank and float bowls (Chapter 4).
5 Intake air leak. Check for loose carburetor-to-intake manifold connections, loose or missing vacuum gauge access port cap or hose, or loose carburetor top (Chapter 4).
6 Idle speed incorrect. Turn idle sped adjuster screw until the engine idles at the specified rpm (Chapters 1 and 4).

9 Rough idle

1 Ignition malfunction. See Chapter 5.
2 Idle speed incorrect. See Chapter 1.
3 Carburetors not synchronized. Adjust carburetors with vacuum gauge set or manometer as outlined in Chapter 1.
4 Carburetor malfunction. See Chapter 4.
5 Fuel contaminated. The fuel can be contaminated with either dirt or water, or can change chemically if the machine is allowed to sit for several months or more. Drain the tank and float bowls. If the problem is severe, a carburetor overhaul may be necessary (Chapters 1 and 4).
6 Intake air leak.
7 Air cleaner clogged. Service or replace air filter element (Chapter 1).

Poor running at low speed

10 Spark weak

1 Battery voltage low. Check and recharge battery (Chapter 8).

2 Spark plug fouled, defective or worn out. Refer to Chapter 1 for spark plug cause.
3 Spark plug cap or high tension wiring defective. Refer to Chapters 1 and 5 for details on the ignition system.
4 Spark plug cap not making contact.
5 Incorrect spark plug. Wrong type, heat range or cap configuration. Check and install correct plugs listed in Chapter 1. A cold plug or one with a recessed firing electrode will not operate at low speeds without fouling.
6 TCI unit defective. See Chapter 5.
7 Pickup coil defective. See Chapter 5.
8 Ignition coil(s) defective. See Chapter 5.

---

11 Fuel/air mixture incorrect

1 Pilot screw(s) out of adjustment (Chapters 1 and 4).
2 Pilot jet or air passage clogged. Remove and overhaul the carburetors (Chapter 4).
3 Air bleed holes clogged. Remove carburetor and blow out all passages (Chapter 4).
4 Air cleaner clogged, poorly sealed or missing.
5 Air cleaner-to-carburetor boot poorly sealed. Look for cracks, holes or loose clamps and replace or repair defective parts.
6 Fuel level too high or too low. Adjust the floats (Chapter 4).
7 Fuel tank air vent obstructed. Make sure that the air vent passage in the filler cap is open.
8 Carburetor intake manifolds loose. Check for cracks, breaks, tears or loose clamps or bolts. Repair or replace the rubber boots.

---

12 Compression low

1 Spark plug loose. Remove the plug and inspect the threads. Reinstall and tighten to the specified torque (Chapter 1).
2 Cylinder head not sufficiently tightened down. If the cylinder head is suspected of being loose, then there's a chance that the gasket and head are damaged if the problem has persisted for any length of time. The head bolts should be tightened to the proper torque in the correct sequence (Chapter 2).
3 Improper valve clearance. This means that the valve is not closing completely and compression pressure is leaking past the valve. Check and adjust the valve clearances (Chapter 1).
4 Cylinder and/or piston worn. Excessive wear will cause compression pressure to leak past the rings. This is usually accompanied by worn rings as well. A top end overhaul is necessary (Chapter 2).
5 Piston rings worn, weak, broken, or sticking. Broken or sticking piston rings usually indicate a lubrication or carburetion problem that causes excess carbon deposits or seizures to form on the pistons and rings. Top end overhaul is necessary (Chapter 2).
6 Piston ring-to-groove clearance excessive. This is caused by excessive wear of the piston ring lands. Piston replacement is necessary (Chapter 2).
7 Cylinder head gasket damaged. If the heads are allowed to become loose, or if excessive carbon build-up on the piston crown and combustion chamber causes extremely high compression, the head gasket may leak. Retorquing the head is not always sufficient to restore the seal, so gasket replacement is necessary (Chapter 2).
8 Cylinder head warped. This is caused by overheating or improperly tightened head bolts. Machine shop resurfacing or head replacement is necessary (Chapter 2).
9 Valve spring broken or weak. Caused by component failure or wear; the spring(s) must be replaced (Chapter 2).
10 Valve not seating properly. This is caused by a bent valve (from over revving or improper valve adjustment), burned valve or seat (improper carburetion) or an accumulation of carbon deposits on the seat (from carburetion, lubrication problems). The valves must be cleaned and/or replaced and the seats serviced if possible (Chapter 2).

---

13 Poor acceleration

1 Accelerator pump defective. Overhaul the carburetors (Chapter 4).
2 Timing not advancing. The pickup coil unit or the alternator may be defective. If so, they must be replaced with new ones, as they cannot be repaired.
3 Carburetors not synchronized. Adjust them with a vacuum gauge set or manometer (Chapter 1).
4 Engine oil viscosity too high. Using a heavier oil than that recommended in Chapter 1 can damage the oil pump or lubrication system and cause drag on the engine.
5 Brakes dragging. Usually caused by debris which has entered the brake piston sealing boot, or from a warped disc or bent axle. Repair as necessary (Chapter 7).

Poor running or no power at high speed

14 Firing incorrect

1 Timing not advancing.
2 Spark plug fouled, defective or worn out. See Chapter 1 for spark plug maintenance.
3 Spark plug cap or high tension wiring defective. See Chapters 1 and 5 for details on the ignition system.
4 Spark plug cap not in good contact. See Chapter 5.
5 Incorrect spark plug. Wrong type, heat range or cap configuration. Check and install correct plugs listed in Chapter 1. A cold plug or one with a recessed firing electrode will not operate at low speeds without fouling.
6 TCI unit defective. See Chapter 5.
7 Ignition coil(s) defective. See Chapter 5.

---

15 Fuel/air mixture incorrect

1 Main jet clogged. Dirt, water and other contaminants can clog the main jets. Clean the fuel petcock filter, the float bowl area, and the jets and carburetor orifices (Chapter 4).
2 Main jet wrong size. The standard jetting is for sea level atmospheric pressure and oxygen content.
3 Throttle shaft-to-carburetor body clearance excessive. Refer to Chapter 4 for inspection and part replacement procedures.
4 Air bleed holes clogged. Remove and overhaul carburetors (Chapter 4).
5 Air cleaner clogged, poorly sealed or missing.
6 Air cleaner-to-carburetor boot poorly sealed. Look for cracks, holes or loose clamps, and replace or repair defective parts.
7 Fuel level too high or too low. Adjust the float(s) (Chapter 4).
8 Fuel tank air vent obstructed. Make sure that the air vent passage in the filler cap is open.
9 Carburetor intake manifolds loose. Check for cracks, breaks, tears or loose clamps or bolts. Repair or replace the rubber boots (Chapter 2).
10 Fuel petcock clogged. Remove the petcock and clean it and the filter (Chapter 1).
11 Fuel line clogged. Pull the fuel line loose and carefully blow through it.
12 Hole or crack in fuel petcock vacuum hose.

---

16 Compression low

1 Spark plug loose. Remove the plug and inspect the threads. Reinstall and tighten to the specified torque (Chapter 1).
2 Cylinder head not sufficiently tightened down. If the cylinder head is suspected of being loose, then there's a chance that the gasket and head are damaged if the problem has persisted for any length of time. The head bolts should be tightened to the proper torque in the correct sequence (Chapter 2).
3 Improper valve clearance. This means that the valve is not closing completely and compression pressure is leaking past the valve. Check and adjust the valve clearances (Chapter 1).
4 Cylinder and/or piston worn. Excessive wear will cause compression pressure to leak past the rings. This is usually accompanied by worn rings as well. A top end overhaul is necessary (Chapter 2).
5 Piston rings worn, weak, broken, or sticking. Broken or sticking piston rings usually indicate a lubrication or carburetion problem that causes excess carbon deposits or seizures to form on the pistons and

rings. Top end overhaul is necessary (Chapter 2).
6   Piston ring-to-groove clearance excessive. This is caused by excessive wear of the piston ring lands. Piston replacement is necessary (Chapter 2).
7   Cylinder head gasket damaged. If the heads are allowed to become loose, or if excessive carbon build-up on the piston crown and combustion chamber causes extremely high compression, the head gasket may leak. Retorquing the head is not always sufficient to restore the seal, so gasket replacement is necessary (Chapter 2).
8   Cylinder head warped. This is caused by overheating or improperly tightened head bolts. Machine shop resurfacing or head replacement is necessary (Chapter 2).
9   Valve spring broken or weak. Caused by component failure or wear; the spring(s) must be replaced (Chapter 2).
10   Valve not seating properly. This is caused by a bent valve (from over revving or improper valve adjustment), burned valve or seat (improper carburetion) or an accumulation of carbon deposits on the seat (from carburetion, lubrication problems). The valves must be cleaned and/or replaced and the seats serviced if possible (Chapter 2).

17   Knocking or pinging

1   Carbon build-up in combustion chamber. Use of a fuel additive that will dissolve the adhesive bonding the carbon particles to the crown and chamber is the easiest way to remove the build-up. Otherwise, the cylinder head will have to be removed and decarbonized (Chapter 2).
2   Incorrect or poor quality fuel. Old or improper grades of gasoline can cause detonation. This causes the piston to rattle, thus the knocking or pinging sound. Drain old gas and always use the recommended fuel grade.
3   Spark plug heat range incorrect. Uncontrolled detonation indicates that the plug heat range is too hot. The plug in effect becomes a glow plug, raising cylinder temperatures. Install the proper heat range plug (Chapter 1).
4   Improper air/fuel mixture. This will cause the cylinder to run hot, which leads to detonation. Clogged jets or an air leak can cause this imbalance. See Chapter 4.

18   Miscellaneous causes

1   Throttle valve doesn't open fully. Adjust the cable slack (Chapter 1).
2   Clutch slipping. Caused by a cable that is improperly adjusted or snagging or damaged, loose or worn clutch components. Refer to Chapters 1 and 2 for adjustment and overhaul procedures.
3   Timing not advancing.
4   Engine oil viscosity too high. Using a heavier oil than the one recommended in Chapter 1 can damage the oil pump or lubrication system and cause drag on the engine.
5   Brakes dragging. Usually caused by debris which has entered the brake piston sealing boot, or from a warped disc or bent axle. Repair as necessary.

Overheating

19   Cooling system not operating properly

1   Coolant level low. Check coolant level as described in Chapter 1. If coolant level is low, the engine will overheat.
2   Leak in cooling system. Check cooling system hoses and radiator for leaks and other damage. Repair or replace parts as necessary (Chapter 3).
3   Thermostat sticking open or closed. Check and replace as described in Chapter 3.
4   Faulty radiator cap. Remove the cap and have it pressure checked at a service station.
5   Coolant passages clogged. Have the entire system drained and flushed, then refill with new coolant.
6   Water pump defective. Remove the pump and check the components.
7   Clogged radiator fins. Clean them by blowing compressed air through the fins from the back side.

20   Firing incorrect

1   Spark plug fouled, defective or worn out. See Chapter 1 for spark plug maintenance.
2   Incorrect spark plug.
3   Improper ignition timing. Timing that is too far advanced will cause high cylinder temperatures and lead to overheating (Chapter 5).

21   Fuel/air mixture incorrect

1   Main jet clogged. Dirt, water and other contaminants can clog the main jets. Clean the fuel petcock filter, the float bowl area and the jets and carburetor orifices (Chapter 4).
2   Main jet wrong size. The standard jetting is for sea level atmospheric pressure and oxygen content.
3   Air cleaner poorly sealed or missing.
4   Air cleaner-to-carburetor boot poorly sealed. Look for cracks, holes or loose clamps and replace or repair.
5   Fuel level too low. Adjust the float(s) (Chapter 4).
6   Fuel tank air vent obstructed. Make sure that the air vent passage in the filler cap is open.
7   Carburetor intake manifolds loose. Check for cracks, breaks, tears or loose clamps or bolts. Repair or replace the rubber boots (Chapter 2).

22   Compression too high

1   Carbon build-up in combustion chamber. Use of a fuel additive that will dissolve the adhesive bonding the carbon particles to the piston crown and chamber is the easiest way to remove the build-up. Otherwise, the cylinder head will have to be removed and decarbonized (Chapter 2).
2   Improperly machined head surface or installation of incorrect gasket during engine assembly. Check Specifications (Chapter 2).

23   Engine load excessive

1   Clutch slipping. Caused by an out of adjustment or snagging cable or damaged, loose or worn clutch components. Refer to Chapters 1 and 2 for adjustment and overhaul procedures.
2   Engine oil level too high. The addition of too much oil will cause pressurization of the crankcase and inefficient engine operation. Check Specifications and drain to proper level (Chapter 1).
3   Engine oil viscosity too high. Using a heavier oil than the one recommended in Chapter 1 can damage the oil pump or lubrication system as well as cause drag on the engine.
4   Brakes dragging. Usually caused by debris which has entered the brake piston sealing boot, or from a warped disc or bent axle. Repair as necessary.

24   Lubrication inadequate

1   Engine oil level too low. Friction caused by intermittent lack of lubrication or from oil that is "overworked" can cause overheating. The oil provides a definite cooling function in the engine. Check the oil level (Chapter 1).
2   Poor quality engine oil or incorrect viscosity or type. Oil is rated not only according to viscosity but also according to type. Some oils are not rated high enough for use in this engine. Check the Specifications section and change to the correct oil (Chapter 1).

25   Miscellaneous causes

Modification to exhaust system. Most aftermarket exhaust systems cause the engine to run leaner, which make them run hotter. When installing an accessory exhaust system, always rejet the carburetors.

Clutch problems

26 Clutch slipping

1 No clutch lever play. Adjust clutch lever free play according to the procedure in Chapter 1.
2 Friction plates worn or warped. Overhaul the clutch assembly (Chapter 2).
3 Steel plates worn or warped (Chapter 2).
4 Clutch springs broken or weak. Old or heat-damaged (from slipping clutch) springs should be replaced with new ones (Chapter 2).
5 Clutch release not adjusted properly. See Chapter 1.
6 Clutch inner cable hanging up. Caused by a frayed cable or kinked outer cable. Replace the cable. Repair of a frayed cable is not advised.
7 Clutch release mechanism defective. Check the shaft, cam, actuating arm and pivot. Replace any defective parts (Chapter 2).
8 Clutch hub or housing unevenly worn. This causes improper engagement of the discs. Replace the damaged or worn parts (Chapter 2).

27 Clutch not disengaging completely

1 Clutch lever play excessive. Adjust at bars or at engine (Chapter 1).
2 Clutch plates warped or damaged. This will cause clutch drag, which in turn causes the machine to creep. Overhaul the clutch assembly (Chapter 2).
3 Clutch spring tension uneven. Usually caused by a sagged or broken spring. Check and replace the springs (Chapter 2).
4 Engine oil deteriorated. Old, thin, worn out oil will not provide proper lubrication for the discs, causing the clutch to drag. Replace the oil and filter (Chapter 1).
5 Engine oil viscosity too high. Using a heavier oil than recommended in Chapter 1 can cause the plates to stick together, putting a drag on the engine. Change to the correct weight oil (Chapter 1).
6 Clutch housing seized on shaft. Lack of lubrication, severe wear or damage can cause the housing to seize on the shaft. Overhaul of the clutch, and perhaps transmission, may be necessary to repair damage (Chapter 2).
7 Clutch release mechanism defective. Worn or damaged release mechanism parts can stick and fail to apply force to the pressure plate. Overhaul the clutch cover components (Chapter 2).
8 Loose clutch hub nut. Causes housing and hub misalignment putting a drag on the engine. Engagement adjustment continually varies. Overhaul the clutch assembly (Chapter 2).

Gear shifting problems

28 Doesn't go into gear or lever doesn't return

1 Clutch not disengaging. See Section 26.
2 Shift fork(s) bent or seized. Often caused by dropping the machine or from lack of lubrication. Overhaul the transmission (Chapter 2).
3 Gear(s) stuck on shaft. Most often caused by a lack of lubrication or excessive wear in transmission bearings and bushings. Overhaul the transmission (Chapter 2).
4 Shift drum binding. Caused by lubrication failure or excessive wear. Replace the drum and bearings (Chapter 2).
5 Shift lever return spring weak or broken (Chapter 2).
6 Shift lever broken. Splines stripped out of lever or shaft, caused by allowing the lever to get loose or from dropping the machine. Replace necessary parts (Chapter 2).
7 Shift mechanism pawl broken or worn. Full engagement and rotary movement of shift drum results. Replace shaft assembly (Chapter 2).
8 Pawl spring broken. Allows pawl to "float", causing sporadic shift operation. Replace spring (Chapter 2).

29 Jumps out of gear

1 Shift fork(s) worn. Overhaul the transmission (Chapter 2).
2 Gear groove(s) worn. Overhaul the transmission (Chapter 2).
3 Gear dogs or dog slots worn or damaged. The gears should be inspected and replaced. No attempt should be made to service the worn parts.

30 Overshifts

1 Pawl spring weak or broken (Chapter 2).
2 Shift drum stopper lever not functioning (Chapter 2).

Abnormal engine noise

31 Knocking or pinging

1 Carbon build-up in combustion chamber. Use of a fuel additive that will dissolve the adhesive bonding the carbon particles to the piston crown and chamber is the easiest way to remove the build-up. Otherwise, the cylinder head will have to be removed and decarbonized (Chapter 2).
2 Incorrect or poor quality fuel. Old or improper fuel can cause detonation. This causes the piston to rattle, thus the knocking or pinging sound. Drain the old gas and always use the recommended grade fuel (Chapter 4).
3 Spark plug heat range incorrect. Uncontrolled detonation indicates that the plug heat range is too hot. The plug in effect becomes a glow plug, raising cylinder temperatures. Install the proper heat range plug (Chapter 1).
4 Improper air/fuel mixture. This will cause the cylinder to run hot and lead to detonation. Clogged jets or an air leak can cause this imbalance. See Chapter 4.

32 Piston slap or rattling

1 Cylinder-to-piston clearance excessive. Caused by improper assembly. Inspect and overhaul top end parts (Chapter 2).
2 Connecting rod bent. Caused by over revving, trying to start a badly flooded engine or from ingesting a foreign object into the combustion chamber. Replace the damaged parts (Chapter 2).
3 Piston pin or piston pin bore worn or seized from wear or lack of lubrication. Replace damaged parts (Chapter 2).
4 Piston ring(s) worn, broken or sticking. Overhaul the top end (Chapter 2).
5 Piston seizure damage. Usually from lack of lubrication or overheating. Replace the pistons and bore the cylinders, as necessary (Chapter 2).
6 Connecting rod big and/or small end clearance excessive. Caused by excessive wear or lack of lubrication. Replace worn parts.

33 Valve noise

1 Incorrect valve clearances. Adjust the clearances by referring to Chapter 1.
2 Valve spring broken or weak. Check and replace weak valve springs (Chapter 2).
3 Camshaft or cylinder head worn or damaged. Lack of lubrication at high rpm is usually the cause of damage. Insufficient oil or failure to change the oil at the recommended intervals are the chief causes. Since there are no replaceable bearings in the head, the head itself will have to be replaced if there is excessive wear or damage (Chapter 2).

34 Other noise

1 Cylinder head gasket leaking. This will cause compression leakage into the cooling system (which may show up as air bubbles in the coolant in the radiator). Also, coolant may get into the oil (which will turn the oil gray). In either case, have the cooling system checked by a dealer service department.
2 Exhaust pipe leaking at cylinder head connection. Caused by improper fit of pipe(s) or loose exhaust flange. All exhaust fasteners should be tightened evenly and carefully. Failure to do this will lead to a leak.
3 Crankshaft runout excessive. Caused by a bent crankshaft (from over revving) or damage from an upper cylinder component failure. Can also be attributed to dropping the machine on either of the crankshaft ends.
4 Engine mounting bolts loose. Tighten all engine mount bolts to the

specified torque (Chapter 2).
5   Crankshaft bearings worn (Chapter 2).
6   Camshaft chain tensioner defective. Replace according to the procedure in Chapter 2.
7   Camshaft chain, sprockets or guides worn (Chapter 2).
8   Loose alternator rotor. Tighten the mounting bolt to the specified torque Chapter 2).

Abnormal driveline noise

35   Clutch noise

1   Clutch housing/friction plate clearance excessive (Chapter 2).
2   Loose or damaged clutch pressure plate and/or bolts (Chapter 2).

36   Transmission noise

1   Bearings worn. Also includes the possibility that the shafts are worn. Overhaul the transmission (Chapter 2).
2   Gears worn or chipped (Chapter 2).
3   Metal chips jammed in gear teeth. Probably pieces from a broken clutch, gear or shift mechanism that were picked up by the gears. This will cause early bearing failure (Chapter 2).
4   Engine oil level too low. Causes a howl from transmission. Also affects engine power and clutch operation (Chapter 1).

37   Driveshaft or final drive noise

Note: *If unusual noises or other problems with the driveshaft or final drive occur, take the motorcycle to a reputable dealer service department for diagnosis and repair.*

Abnormal frame and suspension noise

38   Front end noise

1   Low fluid level or improper viscosity oil. This can sound like ''spurting'' and is usually accompanied by irregular fork action (Chapter 6).
2   Spring weak or broken. Makes a clicking or scraping sound. Fork oil, when drained, will have a lot of metal particles in it (Chapter 6).
3   Steering head bearings loose or damaged. Clicks when braking. Check and adjust or replace as necessary (Chapter 6).
4   Fork clamps loose. Make sure all fork clamp pinch bolts are tight (Chapter 6).
5   Fork tube bent. Good possibility if machine has been dropped. Replace tube with a new one (Chapter 6).
6   Front axle or axle clamp bolt loose. Tighten them to the specified torque (Chapter 6).

39   Shock absorber noise

1   Fluid level incorrect. Indicates a leak caused by defective seal. Shock will be covered with oil. Replace shock (Chapter 6).
2   Defective shock absorber with internal damage. This is in the body of the shock and cannot be remedied. The shock must be replaced with a new one (Chapter 6).
3   Bent or damaged shock body. Replace the shock with a new one (Chapter 6).

40   Disc brake noise

1   Squeal caused by shim not installed or positioned correctly (Chapter 7).
2   Squeal caused by dust on brake pads. Usually found in combination with glazed pads. Clean using brake cleaning solvent (Chapter 7).
3   Contamination of brake pads. Oil, brake fluid or dirt causing brake to chatter or squeal. Clean or replace pads (Chapter 7).
4   Pads glazed. Caused by excessive heat from prolonged use or from contamination. Do not use sandpaper, emery cloth, carborundum cloth or any other abrasive to roughen the pad surfaces as abrasives will stay in the pad material and damage the disc. A very fine flat file can be used, but pad replacement is suggested as a cure (Chapter 7).
5   Disc warped. Can cause a chattering, clicking or intermittent squeal. Usually accompanied by a pulsating lever and uneven braking. Resurface or replace the disc (Chapter 7).

Oil pressure indicator light comes on

41   Engine lubrication system

1   Engine oil pump defective (Chapter 2).
2   Engine oil level low. Inspect for leak or other problem causing low oil level and add recommmended lubricant (Chapters 1 and 2).
3   Engine oil viscosity too low. Very old, thin oil or an improper weight of oil used in engine. Change to correct lubricant (Chapter 1).
4   Camshaft or journals worn. Excessive wear causing drop in oil pressure. Replace cam and/or head. Abnormal wear could be caused by oil starvation at high rpm from low oil level or improper oil weight or type (Chapter 1).
5   Crankshaft and/or bearings worn. Same problems as paragraph 4. Check and replace crankshaft and bearings (Chapter 2).

42   Electrical system

1   Oil pressure switch defective. Check the switch according to the procedure in Chapter 8. Replace it if it is defective.
2   Oil pressure indicator light wiring system defective. Check for pinched, shorted, disconnected or damaged wiring (Chapter 8).

Excessive exhaust smoke

43   White smoke

1   Piston oil ring worn. The ring may be broken or damaged, causing oil from the crankcase to be pulled past the piston into the combustion chamber. Replace the rings with new ones (Chapter 2).
2   Cylinders worn, cracked, or scored. Caused by overheating or oil starvation. The cylinders will have to be rebored and new pistons installed.
3   Valve oil seal damaged or worn. Replace oil seals with new ones (Chapter 2).
4   Valve guide worn. Perform a complete valve job (Chapter 2).
5   Engine oil level too high, which causes oil to be forced past the rings. Drain oil to the proper level (Chapter 1).
6   Head gasket broken between oil return and cylinder. Causes oil to be pulled into combustion chamber. Replace the head gasket and check the head for warpage (Chapter 2).
7   Abnormal crankcase pressurization, which forces oil past the rings. Clogged breather or hoses usually the cause (Chapter 1).

44   Black smoke

1   Air cleaner clogged. Clean or replace the element (Chapter 1).
2   Main jet too large or loose. Compare the jet size to the Specifications (Chapter 4).
3   Accelerator pump check ball stuck.
4   Choke stuck open, causing fuel to be pulled through choke circuit (Chapter 4).
5   Fuel level too high. Check and adjust the float level as necessary (Chapter 4).
6   Inlet needle held off needle seat. Clean float bowl and fuel line and replace needle and seat if necessary (Chapter 4).

45   Brown smoke

1   Main jet too small or clogged. Lean condition caused by wrong size main jet or by a restricted orifice. Clean float bowl and jets and compare jet size to Specifications (Chapter 4).

2 Fuel flow insufficient. Fuel inlet needle valve stuck closed due to chemical reaction with old gas. Float level incorrect. Restricted fuel line. Clean line and float bowl and adjust floats if necessary (Chapter 4).
3 Carburetor intake manifolds loose (Chapter 4).
4 Air cleaner poorly sealed or not installed (Chapter 1).

Poor handling or stability

46   Handlebar hard to turn

1 Steering stem locknut too tight (Chapter 6).
2 Bearings damaged. Roughness can be felt as the bars are turned from side-to-side. Replace bearings and races (Chapter 6).
3 Races dented or worn. Denting results from wear in only one position (i.e., straight ahead) from impacting an immovable object or hole or from dropping the machine. Replace races and bearings (Chapter 6).
4 Steering stem lubrication inadequate. Causes are grease getting hard from age or being washed out by high pressure car washes. Disassemble steering head and repack bearings (Chapter 6).
5 Steering stem bent. Caused by hitting a curb or hole or from dropping the machine. Replace damaged part. Do not try to straighten stem (Chapter 6).
6 Front tire air pressure too low (Chapter 1).

47   Handlebar shakes or vibrates excessively

1 Tires worn or out of balance (Chapter 7).
2 Swingarm bearings worn. Replace worn bearings by referring to Chapter 6.
3 Rim(s) warped or damaged. Inspect wheels for runout (Chapter 7).
4 Wheel bearings worn. Worn front or rear wheel bearings can cause poor tracking. Worn front bearings will cause wobble (Chapter 7).
5 Handlebar clamp bolts loose (Chapter 6).
6 Steering stem or fork clamps loose. Tighten them to the specified torque (Chapter 6).
7 Motor mount bolts loose. Will cause excessive vibration with increased engine rpm (Chapter 2).

48   Handlebar pulls to one side

1 Frame bent. Definitely suspect this if the machine has been dropped. May or may not be accompanied by cracking near the bend. Replace the frame (Chapter 6).
2 Wheel out of alignment. Caused by improper location of axle spacers or from bent steering stem or frame (Chapter 6).
3 Swingarm bent or twisted. Caused by age (metal fatigue) or impact damage. Replace the arm (Chapter 6).
4 Steering stem bent. Caused by impact damage or from dropping the motorcycle. Replace the steering stem (Chapter 6).
5 Fork leg bent. Disassemble the forks and replace the damaged parts (Chapter 6).
6 Fork oil level uneven.

49   Poor shock absorbing qualities

1 Too hard:
 a) Fork oil level excessive (Chapter 6).
 b) Fork oil viscosity too high. Use a lighter oil, per the Specifications (Chapter 6).
 c) Fork tube bent. Causes a harsh, sticking feeling (Chapter 6).
 d) Shock shaft or body bent or damaged (Chapter 6).
 e) Fork internal damage (Chapter 6).
 f) Shock internal damage.
 g) Tire pressure too high (Chapters 1 and 7).
2 Too soft:
 a) Fork or shock oil insufficient and/or leaking (Chapter 6).
 b) Fork oil level too low (Chapter 6).
 c) Fork oil viscosity too light (Chapter 6).
 d) Fork springs weak or broken (Chapter 6).

Braking problems

50   Brakes are spongy, don't hold (hyd. disc brakes only)

1 Air in brake line. Caused by inattention to master cylinder fluid level or by leakage. Locate problem and bleed brakes (Chapter 7).
2 Pad or disc worn (Chapters 1 and 7).
3 Brake fluid leak. See paragraph 1.
4 Contaminated pads. Caused by contamination with oil, grease, brake fluid, etc. Clean or replace pads. Clean disc thoroughly with brake cleaner (Chapter 7).
5 Brake fluid deteriorated. Fluid is old or contaminated. Drain system, replenish with new fluid and bleed the system (Chapter 7).
6 Master cylinder internal parts worn or damaged causing fluid to bypass (Chapter 7).
7 Master cylinder bore scratched. From ingestion of foreign material or broken spring. Repair or replace master cylinder (Chapter 7).
8 Disc warped. Resurface or replace disc (Chapter 7).

51   Brake lever pulsates

1 Disc warped. Resurface or replace disc (Chapter 7).
2 Axle bent. Replace axle (Chapter 6).
3 Brake carrier or caliper bolts loose (Chapter 7).
4 Brake caliper shafts damaged or sticking, causing caliper to bind. Lube the shafts and/or replace them if they are corroded or bent (Chapter 7).
5 Wheel warped or otherwise damaged (Chapter 7).
6 Wheel bearings damaged or worn (Chapter 7).

52   Brakes drag

1 Master cylinder piston seized. Caused by wear or damage to piston or cylinder bore (Chapter 7).
2 Lever balky or stuck. Check pivot and lubricate (Chapter 7).
3 Brake caliper binds. Caused by inadequate lubrication or damage to caliper shafts (Chapter 7).
4 Brake caliper piston seized in bore. Caused by wear or ingestion of dirt past deteriorated seal (Chapter 7).
5 Brake pad damaged. Pad material separating from backing plate. Usually caused by faulty manufacturing process or from contact with chemicals. Replace pads (Chapter 7).
6 Pads improperly installed (Chapter 7).
7 Rear brake pedal free play insufficient.

Electrical problems

53   Battery dead or weak

1 Battery faulty. Caused by sulphated plates which are shorted through the sedimentation or low electrolyte level. Also, broken battery terminal making only occasional contact (Chapter 8).
2 Battery cables making poor contact (Chapter 8).
3 Load excessive. Caused by addition of high wattage lights or other electrical accessories.
4 Ignition switch defective. Switch either grounds internally or fails to shut off system. Replace the switch (Chapter 8).
5 Regulator/rectifier defective (Chapter 8).
6 Stator coil open or shorted (Chapter 8).
7 Wiring faulty. Wiring grounded or connections loose in ignition, charging or lighting circuits (Chapter 8).

54   Battery overcharged

1 Regulator/rectifier defective. Overcharging is noticed when battery gets excessively warm or ''boils'' over (Chapter 8).
2 Battery defective. Replace battery with a new one (Chapter 8).
3 Battery amperage too low, wrong type or size. Install manufacturer's specified amp-hour battery to handle charging load (Chapter 8).

Chapter 1 Tune-up and routine maintenance

Contents

Specifications

Engine

Spark plugs	
Type – US models	NGK D8EA or ND X24ES-U
Type – UK models	NGK DR8ES-L or ND X24ESR-U
Gap	0.024 to 0.028 in (0.6 to 0.7 mm)
Engine idle speed	
US models	1300 ± 50 rpm
UK models	1100 ± 50 rpm
Valve clearances (COLD engine)	
Intake	0.0043 to 0.0059 in (0.11 to 0.15 mm)
Exhaust	0.0063 to 0.0079 in (0.16 to 0.20 mm)
Cylinder compression pressure	
Standard	142 psi
Minimum	128 psi
Maximum	156 psi
Maximum difference between cylinders	14 psi
Cylinder vacuum (at idle)	Above 7.09 in (180 mm) Hg
Vacuum difference between cylinders	Less than 0.390 in (10 mm) Hg

Free play adjustments

Front brake lever free play	0.2 to 0.3 in (5 to 8 mm)
Rear brake pedal free play	0.6 to 1.0 in (15 to 25 mm)
Clutch lever free play	0.08 to 0.12 in (2 to 3 mm)

Miscellaneous

Battery electrolyte specific gravity	1.260 minimum
Minimum tire tread depth	0.030 in (0.8 mm)

Tire pressures (cold)

	Front	Rear
0 – 198 lb (0 – 90 kg) load	26 psi (1.8 kg/cm²)	28 psi (2.0 kg/cm²)
198 – 309 lb (90 – 140 kg) load	28 psi (2.0 kg/cm²)	33 psi (2.3 kg/cm²)
309 – 430 lb (140 – 195 kg) max load	33 psi (2.3 kg/cm²)	40 psi (2.8 kg/cm²)
High speed riding	28 psi (2.0 kg/cm²)	33 psi (2.3 kg/cm²)

Note: loads given apply to total weight of rider, passenger and any accessories or luggage

Torque specifications

	Ft-lb	Nm
Oil drain plug	31	43
Water pump cover drain bolts	7.2	10
Spark plugs	14	20
Cylinder head cover bolts	7.2	10
Exhaust pipe flange bolts	7.2	10
Muffler clamp bolts	14	20
Right side frame tube bolts	19	26
Fork tube pinch bolts	14	20

Recommended lubricants and fluids

Engine/transmission oil
  Type .... Yamalube 4-cycle oil or equivalent grade SE
  Viscosity
    Above 40-degrees F (5-degrees C) .... SAE 20W40
    Below 60-degrees F (15-degrees C) .... SAE 10W30
  Capacity
    With filter change .... 2.9 US qt (2.7 liters)
    Oil change only .... 2.5 US qt (2.4 liters)
Final drive gear oil
  Type .... API GL-4 hypoid gear oil
  Viscosity .... SAE 80 or 80W90
  Capacity .... 0.21 US qt (0.2 liter)
Coolant
  Type .... 50/50 mixture of ethylene glycol based antifreeze and soft water
  Capacity
    Total .... 2.4 US qt (2.3 liter)
    Reservoir only .... 0.4 US qt (0.35 liter)
Brake fluid .... DOT 3
Fork oil – US models
  RJ models
    Type .... 15 weight fork oil
    Amount .... 8.7 oz (257 cc's) per side
    Oil level .... 6.0 in (152 mm) from top edge of fork tube
                                                        (forks completely compressed – no springs)

  RK models
    Type .... 15 weight fork oil
    Amount .... 9.0 oz (267 cc's) per side
    Oil level .... 5.0 in (127 mm) from top edge of fork tube
                                                        (forks completely compressed – no springs)

Fork oil – UK models
  Type .... SAE 10W/30 SE
  Amount .... 8.76 Imp fl oz (249 cc) per side
  Oil level .... 100 – 140 mm (3.9 – 5.5 in)
Miscellaneous
  Wheel bearings .... Medium weight, lithium-based multi-purpose grease
  Swingarm pivot bearings .... Medium weight, lithium-based multi-purpose grease
  Cables and lever pivots .... Yamaha chain and cable lubricant or 10W30 motor oil
  Sidestand/centerstand pivots .... Yamaha chain and cable lubricant, 10W30 motor oil or dry film lubricant
  Brake pedal/shift lever pivots .... Yamaha chain and cable lubricant, 10W30 motor oil or dry film lubricant
  Throttle grip .... Lightweight grease or dry film lubricant

---

1   Introduction to tune-up and routine maintenance

This Chapter covers in detail the checks and procedures necessary for the tune-up and routine maintenance of your motorcycle. It is divided into two parts. The first contains applicable specifications and service data and outlines routine maintenance intervals. The second covers procedures, or how to perform each of the maintenance functions.

Since routine maintenance plays such an important role in the safe and efficient operation of your motorcycle, it is presented here as a comprehensive check list. For the rider who does all his own maintenance, these lists outline the procedures and checks that should be done on a routine basis.

Deciding where to start or plug into the routine maintenance schedule depends on several factors. If you have a motorcycle whose warranty has recently expired, and if it has been maintained according to the warranty standards, you may want to pick up routine maintenance as it coincides with the next mileage or calendar interval. If you have owned the machine for some time but have never performed any maintenance on it, then you may want to start at the nearest interval and include some additional procedures to ensure that nothing important is overlooked. If you have just had a major engine overhaul, then you may want to start the maintenance routine from the beginning. If you have a used machine and have no knowledge of its history or maintenance record, you may desire to combine all the checks into one large service initially and then settle into the maintenance schedule prescribed.

Note that the procedures normally associated with ignition and carburetion tune-ups are included in the routine maintenance schedule. A regular tune-up ensures optimum engine performance and helps prevent engine damage due to improper carburetion and ignition timing.

The Sections which actually outline the inspection and maintenance procedures are written as step-by-step comprehensive guides to the actual performance of the work. They explain in detail each of the routine inspections and maintenance procedures on the check list. References to additional information in applicable Chapters is also included and should not be overlooked. **Note:** *Before beginning any tune-up or maintenance procedures on an RK model, refer to Chapter 9 to see which portions of the fairing, if any, must be removed.*

Before beginning any actual maintenance or repair, the machine should be cleaned thoroughly, especially around the oil filter housing, spark plugs, cylinder head covers, side covers, carburetors, etc. Cleaning will help ensure that dirt does not contaminate the engine and will allow you to detect wear and damage that could otherwise easily go unnoticed.

---

2   Routine maintenance intervals

Note: *The pre-operation checks outlined in the owner's manual cover checks and maintenance that should be carried out on a daily or preride basis. It is condensed and included here to remind you of its importance. Always perform the pre-ride inspection at every maintenance interval (in addition to the procedures listed). The following intervals are recommended by the motorcycle manufacturer, not the publisher of this manual. It may be wise, in many cases, to perform the maintenance at shorter intervals.*

Daily or before riding
Check the engine oil level
Check the engine coolant level and look for obvious leaks
Check the fuel level in the tank and look for fuel leaks
Check the operation of both brakes – look for fluid leakage
   (hydraulic disc brakes only) and adjust brake free play if
   necessary
Check the tires for damage, the presence of foreign objects and
   correct air pressure
Check the final drive gear case for leaks
Check the throttle for smooth operation and correct free play
Check the clutch operation
Check the battery electrolyte level
Check for proper operation of the headlight, taillight, brake light,
   turn signals, indicator lights and horn
Make sure the engine STOP switch works properly

At 600 miles or 1 month
Change the engine oil and filter
Clean the air filter element
Adjust clutch lever free play
Change the final drive gear oil
Lubricate all cables

At 3000 miles or 7 months
Check and adjust the valve clearances
Clean the spark plugs and check/adjust the gaps
Check the crankcase ventilation system hose for cracks and other
   damage
Check the fuel lines and vacuum hose for cracks and other
   damage
Check the exhaust system joints for leaks
Check/adjust carburetor synchronization
Check/adjust idle speed and throttle free play
Change the engine oil and filter
Clean the air filter element
Adjust brake pedal free play
Adjust clutch lever free play
Lubricate the clutch and brake lever pivots
Lubricate all cables
Lubricate the shift lever/brake pedal pivots and the sidestand/cen-
   terstand pivots
Check the steering head bearings
Check the wheel bearings
Check the battery electrolyte specific gravity and make sure the
   vent tube is routed properly

Every 2500 miles or 6 months
Clean the spark plugs and check/adjust the gaps
Check the exhaust system joints for leaks
Check/adjust carburetor synchronization
Check/adjust idle speed and throttle free play
Change the engine oil and filter
Check the brakes – replace the pads and shoes with new ones as
   required
Adjust clutch lever free play
Lubricate the clutch and brake lever pivots
Lubricate all cables
Lubricate the shift lever/brake pedal pivots and the sidestand/cen-
   terstand pivots
Check the steering head bearings
Check the wheel bearings
Check the battery electrolyte specific gravity and make sure the
   vent tube is routed properly

Every 5000 miles or 12 months
Check and adjust the valve clearances
Check the crankcase ventilation system
Check the fuel system
Clean the air filter element
Change the final drive gear oil

Every 7500 miles or 18 months
Replace the spark plugs with new ones

Every 10000 miles or 24 months
Check the cooling system and replace the coolant
Repack the steering head bearings
Change the fork oil

---

3   Engine oil level — check

1   Place the motorcycle on the centerstand, then start the engine and
allow it to reach normal operating temperature. **Caution:** *Do not run
the engine in an enclosed space such as a garage or shop.*
2   Stop the engine and allow the machine to sit undisturbed on the
centerstand for about five minutes.
3   With the engine off, check the oil level in the window located at
the lower part of the left crankcase cover. The oil level should be between the Maximum and Minimum level marks on the window (photo).
4   If the level is below the Minimum mark, remove the oil filler cap
from the front of the left crankcase cover (photo) and add enough oil
of the recommended grade and type to bring the level up to the Maximum mark. Do not overfill.

---

4   Brake fluid level — check

1   In order to ensure proper operation of the hydraulic front disc brake,
the fluid level in the master cylinder must be properly maintained.
2   With the motorcycle on the centerstand, turn the handlebars until
the top of the master cylinder is as level as possible. If necessary, loosen

3.3   The engine oil level must be between the Minimum and
Maximum marks (arrows) at the window

3.4 Engine oil filler cap location

the brake lever clamp bolts and rotate the master cylinder assembly slightly to make it level.

3   Look closely at the inspection window in the master cylinder reservoir. Make sure that the fluid level is above the Lower mark on the reservoir (photo).

4   If the level is low, the fluid must be replenished. Before removing the master cylinder cap, cover the gas tank to protect it from brake fluid spills (which will damage the paint) and remove all dust and dirt from the area around the cap.

5   Remove the screws and lift off the cap and rubber diaphragm. **Note:** *Do not operate the brake lever with the cap removed.*

6   Add new, clean brake fluid of the recommended type until the level is above the inspection window. Do not mix different brands of brake fluid in the reservoir, as they may not be compatible.

7   Replace the rubber diaphragm and the cover. Tighten the screws evenly, but do not overtighten them.

8   If the brake fluid level was low, check the entire system for leaks.

9   Wipe any spilled fluid off the reservoir body and reposition and tighten the brake lever and master cylinder assembly if it was moved.

5   Coolant level — check

1   The engine must be cold for the results to be accurate, so always perform this check before starting the engine for the first time each day. Remove the right side cover and locate the coolant reservoir.

2   The coolant level is satisfactory if it is between the Low and Full marks on the reservoir (photo). If the level is at or below the Low mark, add *soft water* until the Full level is reached. If the coolant level seems to be consistently low, check the entire cooling system for leaks.

6   Battery electrolyte level/specific gravity — check

Caution: *Be extremely careful when handling or working around the battery. The electrolyte is very caustic and an explosive gas is given off when the battery is charging.*

1   To check and replenish the battery electrolyte, it will be necessary to remove the left side cover.

2   With the side cover removed, locate the inspection window in the upper left corner of the plastic battery cover. The electrolyte level in the front cell can be checked in the window — it should be between the upper and lower level marks (photo). To check the remaining cells, remove the plastic cover (it is held in place with two screws). The electrolyte level in each cell should be between the upper and lower marks on the battery case (photo).

3   If it is low, pull the battery out slightly, remove the cell caps and fill each cell to the upper level mark with distilled water. Do not use

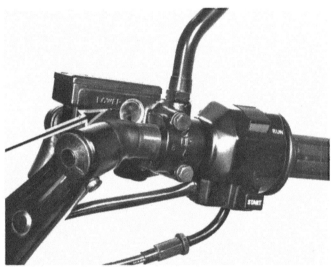

4.3 The brake fluid level must be above the Lower mark (arrow) at the inspection window in the reservoir

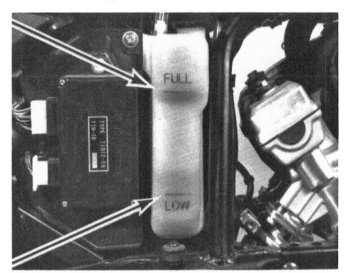

5.2 The coolant level must be between the Low and Full marks (arrows) on the reservoir

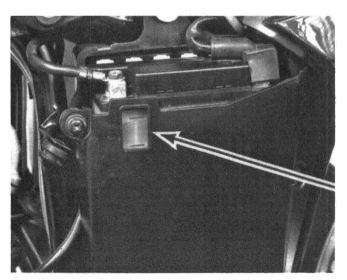

6.2a The battery electrolyte level can be checked at a glance by looking through the inspection window (arrow) in the cover

6.2b The electrolyte level in all cells must be between the Upper and Lower level marks (arrows)

tap water (except in an emergency) and do not overfill. The cell holes are quite small, so it may help to use a plastic squeeze bottle with a small spout to add the water. If the level is within the marks on the case, additional water is not necessary.

4   Next, check the specific gravity of the electrolyte in each cell with a small hydrometer made especially for motorcycle batteries. These are available from most dealer parts departments or motorcycle accessory stores.

5   Remove the caps, draw some electrolyte from the first cell into the hydrometer (photo) and note the specific gravity. Compare the reading to the Specifications. Return the electrolyte to the appropriate cell and repeat the check for the remaining cells. When the check is complete, rinse the hydrometer thoroughly with clean water.

6   If the specific gravity of the electrolyte in each cell is as specified, the battery is in good condition and is apparently being charged by the machine's charging system.

7   If the specific gravity is low, the battery is not fully charged. This may be due to corroded battery terminals, a dirty battery case, a malfunctioning charging system, or loose or corroded wiring connections. On the other hand, it may be that the battery is worn out, especially if the machine is old, or that infrequent use of the motorcycle prevents normal charging from taking place.

8   Be sure to correct any problems and charge the battery if necessary. Refer to Chapter 8, *Electrical system*, for additional battery maintenance

and charging procedures.

9   Install the battery cell caps, the plastic cover and the side cover. Be very careful not to pinch or otherwise restrict the battery vent tube, as the battery may build up enough internal pressure during normal charging system operation to explode.

7 Final drive gear oil level — check

1   Place the motorcycle on the centerstand on a level surface. The engine must be completely cool when this check is made.

2   Clean the area around the final drive assembly oil filler cap (photo), then remove the cap. The oil level should be at the upper outside edge of the threaded hole as shown in the accompanying illustration. If the level is low, add enough oil of the recommended grade and type to bring it up to the specified point.

3   Clean the filler cap, then reinstall and tighten it securely.

8 Brake shoes/pads — wear check

1   The rear brake shoes and front brake pads should be checked at the recommended intervals and replaced with new ones when worn beyond the specified limits.

6.5 Checking electrolyte specific gravity with a special motorcycle battery hydrometer

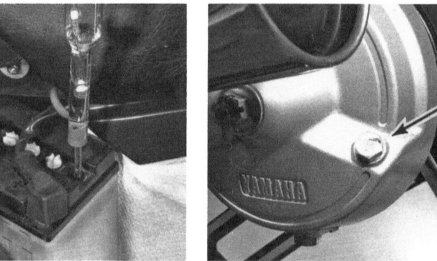

7.2 Final drive gear oil filler cap location

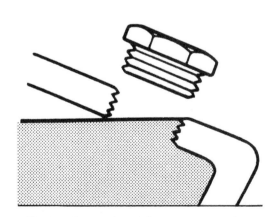

Fig. 1.1   Correct final drive gear oil level (Sec 7)

8.2 Check the front disc brake pads for wear by examining the pad lining material (1) and the wear indicators (2)

Front brake pads

2   To check the disc brake pads, position the front wheel so you can see clearly into the rear of the brake caliper. The brake pads are visible from this angle and should have at least 1/8 inch of lining material remaining on the metal backing plate (photo).

3   If the pads are worn excessively, the wear indicators (small tabs which are an integral part of the metal backing plate) will be touching or nearly touching the brake disc. If the pads are worn to this point, they must be replaced with new ones (refer to Chapter 7).

Rear brake shoes

4   The rear drum brake shoes are checked by applying the brake firmly and checking to see whether the pointer on the brake arm reaches the rear of the wear line on the backing plate (photo).

5   If the pointer reaches the rear of the wear line, the brake shoes are excessively worn and must be replaced (refer to Chapter 7).

9   Brake system — general check

1   A routine general check of the brakes will ensure that any problems are discovered and remedied before the rider's safety is jeopardized.

2   Check the rear brake shoes for excessive wear and the lever, brake rod and pedal for loose connections, excessive play, bends, and other damage. Replace any damaged parts with new ones (Chapter 7).

3   Inspect the tension bar for loose connections and damage. Make sure that the cotter key is properly installed.

4   Check the front brake lever free play and make sure all brake fasteners are tight. Check the brake pads for wear and make sure the fluid level in the reservoir is correct. Look for leaks at the hose connections and check for cracks in the hoses. If the lever is spongy, bleed the brakes as described in Chapter 7.

5   Make sure that the brake light operates when the brake lever and pedal are depressed.

10   Brake light switches — check and adjustment

1   Make sure the brake light is activated when the rear brake pedal is depressed a maximum of 3/4 inch.

2   If adjustment is necessary, hold the switch and turn the adjusting nut on the switch body (photo) until the brake light is activated when required. Turning the nut clockwise will cause the brake light to come on sooner, while turning it counterclockwise will cause it to come on later.

3   The front brake light switch is not adjustable. If it fails to operate properly, replace it with a new one (see Chapter 7, Section 15).

8.4 Check the rear drum brake shoes for wear by applying the brake and checking to see if the pointer (1) reaches the rear of the wear line (2)

10.2 Brake light switch (1) and adjusting nut (2) location

11.2 Rear brake pedal height adjusting screw (1) and locknut (2) location

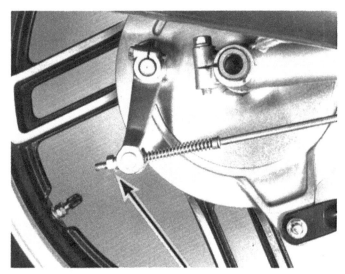

11.6 Brake rod adjusting nut location

11 Rear brake pedal/front brake lever — check and adjustment

Rear brake pedal height
1 Rear brake pedal height is largely a matter of personal preference. Locate the pedal so that the rear brake can be engaged quickly and easily without excessive foot movement.
2 To adjust the height, loosen the locknut on the adjusting screw (photo), turn the screw in or out until the pedal height is correct for you, then retighten the locknut.
3 After the pedal height has been adjusted, always check and adjust the free play.

Rear brake pedal free play
4 Correct brake pedal free play is important for proper brake operation. If there is not enough free play, the brake will drag. Excessive free play may prevent the full application of the brake, which is obviously dangerous.
5 Depress the brake pedal and measure how far the end of the pedal travels before the shoes contact the drum. Compare the results to the Specifications.
6 If adjustment is necessary, turn the adjusting nut on the end of the brake rod (photo) until the proper free play is obtained. Turning the nut clockwise will decrease the play, while turning it counterclockwise will increase the play.

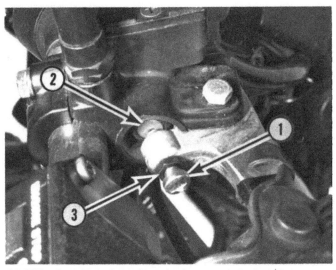

11.8 The front brake lever free play is measured when the adjusting screw (1) contacts the master cylinder piston (2); adjustment requires loosening of the locknut (3)

12.1 Fuel line (1), vacuum hose (2) and petcock mounting screw (arrows) locations

Front brake lever
7 Correct lever free play is necessary for proper brake operation. If there is not enough free play, the brake may drag. Excessive free play may hinder full application of the brake, which is obviously dangerous.
8 Pull back the rubber boot, slowly squeeze the front brake lever and measure how far the end of the lever travels before the end of the adjusting screw contacts the master cylinder piston; this distance is the free play. If adjustment is required, loosen the locknut on the adjusting screw (photo) and turn the screw in or out to obtain the correct free play. Be sure to retighten the locknut after the adjustment is made.
9 **Caution:** *A soft or spongy feeling at the brake lever can indicate the presence of air in the brake system, which will adversely affect the operation of the brake. Refer to Chapter 7 and bleed the brake before operating the motorcycle.*

12 Fuel system — check

1 Check the fuel tank, the petcock, the lines and the carburetors for leaks and evidence of damage (photo).
2 If carburetor gaskets are leaking, the carburetors should be disassembled and rebuilt by referring to Chapter 4.
3 If the fuel petcock is leaking, tightening the screws may help. If leakage persists, the petcock should be disassembled and repaired or replaced with a new one.
4 If the fuel lines are cracked or otherwise deteriorated, replace them with new ones.
5 Check the vacuum hose connected to the petcock. If it is cracked or otherwise damaged, replace it with a new one.

13 Fuel petcock/filter — servicing

Warning: *Gasoline is extremely flammable and extra precautions must be taken when working on any part of the fuel system. Do not smoke and do not allow open flames or bare light bulbs near the motorcycle. Also, do not perform this operation in a garage where a natural gas type water heater is installed.*
1 The fuel filter, which is attached to the fuel petcock, may become clogged and should be removed and cleaned periodically. In order to clean the filter, the fuel tank must be drained and the petcock removed. This can be accomplished with the tank on the motorcycle, but it may not drain completely unless it is removed. **Note:** *There is a drain screw on the petcock, but a container must be held under the opening as the fuel is draining. As a result, the method outlined here is easier and safer.*
2 With the petcock in the On or Reserve position, slide back the hose clamp and pull the fuel line off of the petcock.

13.6 Clean the fuel filter (1) and check the mounting flange O-ring (2) when servicing the fuel petcock

3   Raise the seat, remove the bolt that attaches the rear of the tank to the frame and lift it off. Be careful not to scratch the paint when handling the tank. Attach a separate section of fuel line to the petcock (it should be at least 18 inches long).

4   Support the tank, place the free end of the fuel line in an approved gasoline container, move the petcock lever to the Prime position and drain the fuel out of the tank.

5   Once the tank is emptied, loosen and remove the screws that attach the petcock to the tank. Remove the petcock and filter.

6   Clean the filter (photo) with solvent and blow it dry with compressed air. If the filter is torn or otherwise damaged, replace the entire petcock with a new one. Check the mounting flange O-ring and the gaskets on the screws. If they are damaged, replace them with new ones.

7   Install the O-ring, filter and petcock on the tank, replace the tank and seat and hook up the fuel line. Refill the tank and check carefully for leaks around the mounting flange and screws.

8   If you choose not to remove the tank, make sure the motorcycle is on the centerstand. Place the petcock lever in the On or Reserve position, slide back the hose clamp and pull the fuel line off. Attach a separate section of fuel line to the petcock. Place the free end of the fuel line in an approved gasoline container, move the petcock lever to the Prime position and drain the fuel out of the tank. Loosen and remove the screws, then withdraw the petcock and fuel filter. Be prepared to spill some gasoline, as the tank probably will not drain completely.

14   Crankcase ventilation system — check

The only external components of the crankcase ventilation system, and the only ones requiring inspection, are the rubber hoses leading from each cylinder head cover to the airbox. Carefully examine each hose along its entire length, looking for cracks and other damage. Disconnect the hoses at the covers and see if air will pass through them. If the hoses are damaged or clogged, replace them with new ones.

15   Exhaust system — check

1   Periodically check all of the exhaust system joints for leaks and loose fasteners. If tightening the clamp bolts fails to stop any leaks, replace the gaskets with new ones (a procedure which requires disassembly of the system).

2   The exhaust pipe flange bolts at the cylinder heads (photo) are especially prone to loosening, which could cause damage to the head. Check them frequently and keep them tight.

16   Clutch — check and adjustment

1   Correct clutch free play is necessary to ensure proper clutch opera-

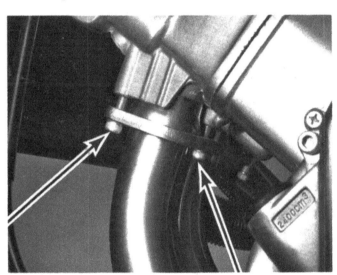

15.2 The exhaust pipe flange bolts (arrows) should be checked frequently and tightened when necessary

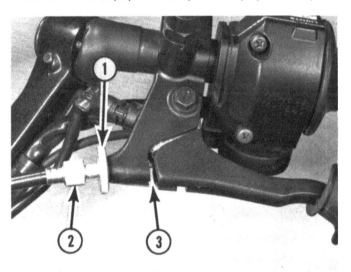

16.2 Loosen the lock wheel (1) and turn the adjuster (2) to obtain the specified gap (3) between the clutch lever and bracket

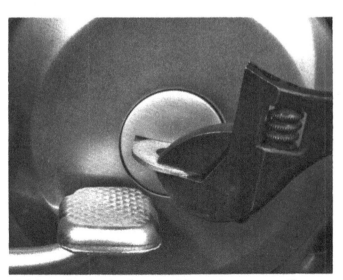

16.5 Removing the plug from the right crankcase cover with a large washer and an adjustable wrench

16.6 Loosen the locknut (1) and turn the adjusting screw (2) to adjust the clutch release mechanism

tion and reasonable clutch service life. Free play normally changes because of cable stretch and clutch wear, so it should be checked and adjusted periodically.

2   Clutch cable free play is checked at the lever on the handlebar. Slowly pull in on the lever until resistance is felt, then note how far the lever has moved away from its bracket at the pivot end. Generally speaking, there should be a gap of approximately 1/16 to 1/8 inch between the lever and the bracket (photo) when resistance from the clutch release mechanism is felt.

3   Free play adjustments can be made at the clutch lever by pulling back the rubber shroud, loosening the lock wheel and turning the adjuster until the desired free play is obtained. Always retighten the lock wheel once the adjustment is complete. If the lever adjuster reaches the end of its travel, a new cable is required.

4   To adjust the clutch release mechanism, loosen the lever adjuster lock wheel and back it off completely. Thread the adjuster into the bracket to produce as much slack as possible in the cable.

5   Remove the plug from the right crankcase cover. To avoid damage to the plug, a large washer and adjustable wrench should be used in place of a screwdriver (photo).

6   The clutch cable is connected to a lever on the bottom of the left side of the engine. Push the lever toward the front of the machine until it stops and see if the pointer on the end of the lever is aligned with the index mark on the crankcase. If it is, the release mechanism does not require adjustment. If it is not, loosen the locknut and turn the adjusting screw in the right crankcase cover (photo) until the pointer and index mark are aligned, then tighten the locknut while holding the adjusting screw stationary with a screwdriver.

7   Reinstall the plug in the right crankcase cover and readjust the clutch lever free play at the handlebar adjuster.

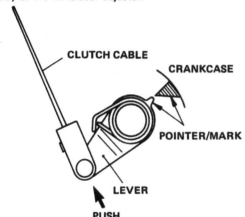

Fig. 1.2   Clutch release mechanism adjustment details   (Sec 16)

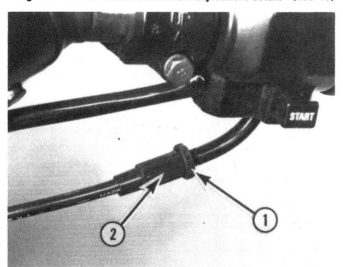

19.3 Loosen the lock wheel (1) and turn the adjuster (2) to change the throttle cable/grip free play

17   Tires/wheels — general check

1   Routine tire and wheel checks should be made with the realization that your safety depends to a great extent on their condition.

2   Check the tires carefully for cuts, tears, embedded nails or other sharp objects and excessive wear. Operation of the motorcycle with excessively worn tires is extremely hazardous, as traction and handling are directly affected. Measure the tread depth at the center of the tire and replace worn tires with new ones when the tread depth is less than specified.

3   Repair or replace punctured tires as soon as damage is noted. Do not try to patch a torn tire, as wheel balance and tire reliability may be impaired.

4   Check the tire pressures when the tires are cold and keep them properly inflated. Proper air pressure will increase tire life and provide maximum stability and ride comfort. Keep in mind that low tire pressures may cause the tire to slip on the rim or come off, while high tire pressures will cause abnormal tread wear and unsafe handling.

5   The cast wheels used on this machine are virtually maintenance free, but they should be kept clean and checked periodically for cracks and other damage. Never attempt to repair damaged cast wheels; they must be replaced with new ones.

6   Check the valve stem locknuts to make sure they are tight. Also, make sure the valve stem cap is in place and tight. If it is missing, install a new one made of metal or hard plastic.

18   Fasteners — check

1   Since vibration of the machine tends to loosen fasteners, all nuts, bolts, screws, etc. should be periodically checked for proper tightness.

2   Pay particular attention to the following:
   Spark plugs
   Engine oil drain plug
   Oil filter cover bolts
   Gearshift lever
   Engine mount bolts
   Shock absorber mount bolts
   Front axle clamp bolt
   Rear axle nut
   Final drive housing drain plug

3   If a torque wrench is available, use it along with the Torque specifications at the beginning of this, or other, Chapters.

19   Throttle operation/grip free play — check and adjustment

1   Make sure the throttle grip rotates easily from fully closed to fully open with the front wheel turned at various angles. The grip should return automatically from fully open to fully closed when released. If the throttle sticks, check the throttle cable for cracks or kinks in the housing. Also, make sure the inner cable is clean and well-lubricated.

2   Check for a small amount of free play at the grip.

3   Free play adjustments can be made at the throttle end of the cable. Loosen the lock wheel on the cable (photo) and turn the adjuster until the desired free play is obtained, then retighten the lock wheel.

20   Headlight aim — check and adjustment

1   An improperly adjusted headlight may cause problems for oncoming traffic or provide poor, unsafe illumination of the road ahead. Before adjusting the headlight, be sure to consult local traffic laws and regulations.

2   The headlight beam can be adjusted both vertically and horizontally (photo). To adjust the headlight vertically, turn the lower adjusting screw as required to raise or lower the beam. Directing the light at the side of a building or other flat, vertical surface will help obtain the proper beam level. Be sure to have someone sit on the machine when making the adjustment.

3   To adjust the light horizontally, turn the upper adjusting screw until the beam is properly oriented.

21 Steering head bearings — check and adjustment

1   The Yamaha Vision is equipped with ball-and-cone type steering head bearings which generally become dented, rough or loose during normal use of the machine. In extreme cases, worn or loose steering head bearings can cause steering wobble that is potentially dangerous.
2   To check the bearings, place the motorcycle on the centerstand and block the machine so the front wheel is in the air.
3   Point the wheel straight ahead and slowly move the handlebars from side-to-side. Dents or roughness in the bearing races will be felt and the bars will not move smoothly.
4   Next, grasp the fork legs and try to move the wheel forward and backward (photo). Any looseness in the steering head bearings will be felt. If play is felt in the bearings, adjust the steering head as follows:
5   Loosen the upper fork pinch bolts and the steering stem bolt (photo). Use a spanner wrench to loosen the upper steering stem ring nut (this nut serves as a jam nut) (photo).
6   Carefully tighten the lower steering stem ring nut until the steering head is tight but does not bind when the forks are turned from side-to-side.
7   Retighten the upper steering stem nut, the steering stem bolt and the upper fork pinch bolts in that order.

8   Recheck the steering head bearings for play as described above. If necessary, repeat the adjustment procedure.
9   Refer to Chapter 7 for steering head bearing lubrication and replacement procedures.

22 Cooling system — check

Note: *Refer to Section 5 and check the coolant level before performing this check.*

1   The entire cooling system should be checked carefully at the recommended intervals. Look for evidence of leaks, check the condition of the coolant, check the radiator for clogged fins and damage and make sure the fan operates when required.
2   Examine each of the rubber coolant hoses along its entire length. Look for cracks, abrasions and other damage. Squeeze each hose at various points. They should feel firm, yet pliable, and return to their original shape when released. If they are dried out or hard, replace them with new ones.
3   Check for evidence of leaks at each cooling system joint. Tighten the hose clamps carefully to prevent future leaks.
4   Remove the plastic radiator cover (it is held in place with four screws) and check the radiator for evidence of leaks and other damage. Leaks

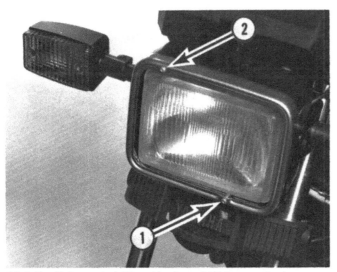

20.2 Headlight vertical (1) and horizontal (2) beam adjusting screws

21.4 Checking the steering head bearings for play

21.5a Upper fork pinch bolts (1), steering stem bolt (2) and upper and lower steering stem ring nuts (3)

21.5b Loosening the upper steering stem ring nut with a spanner wrench

in the radiator leave telltale scale deposits or coolant stains on the outside of the core below the leak. If leaks are noted, remove the radiator (refer to Chapter 3) and have it repaired at a radiator shop or replace it with a new one. **Caution:** *Do not use a liquid leak stopping compound to try to repair leaks.*

5   Check the radiator fins for mud, dirt and insects, which may impede the flow of air through the radiator. If the fins are dirty, force water or low pressure compressed air through the fins from the backside. If the fins are bent or distorted, straighten them carefully with a screwdriver.

6   Remove the radiator cap and check the condition of the coolant in the radiator. If it is rust colored or if accumulations of scale are visible in the radiator, drain, flush and refill the system with new coolant. Check the cap gaskets for cracks and other damage. Have the cap tested by a dealer service department or replace it with a new one. Replace the cap and install the radiator cover.

7   Start the engine and let it reach normal operating temperature, then check for leaks again. As the coolant temperature increases, the fan should come on automatically and the temperature should begin to drop. If it does not, refer to Chapter 8 and check the fan and wiring carefully.

8   If the coolant level is consistently low, and no evidence of leaks can be found, have the entire system pressure checked by a Yamaha dealer service-department, motorcycle repair shop or service station.

23   Cooling system — draining, flushing and refilling

Caution: *Allow the engine to cool completely before performing this maintenance operation.*

1   Remove the plastic radiator cover (it is held in place with four screws) and loosen the radiator cap. Place a large, clean drain pan under the front of the engine.

2   Remove the drain bolt from the bottom of the water pump cover (photo) and allow the coolant to drain into the pan. **Note:** *The coolant will rush out with considerable force, so position the drain pan accordingly.* Remove the radiator cap completely to ensure that all of the coolant can drain.

3   Remove the drain bolts from each cylinder (photos) and allow the coolant to drain into the pan.

4   To drain the radiator, remove the thermostat cover screws, separate the cover from the engine and allow the coolant to drain through the radiator hose into the pan. **Note:** *The coolant will rush out of the hose with some force, so be prepared to catch it in the drain pan.*

5   Flush the system with clean tap water by inserting a garden hose in the radiator filler neck. Allow the water to run through the system until it is clear when it exits the radiator hose and the drain holes. If the radiator is extremely corroded, remove it by referring to Chapter

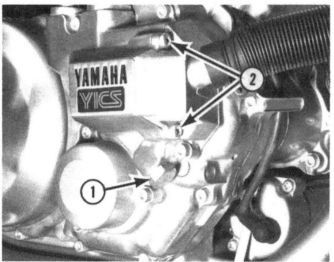

23.2 Water pump drain bolt (1) and thermostat cover mounting screws (2)

23.3a Front cylinder coolant drain bolt location

23.3b Rear cylinder coolant drain bolt location

23.6 Check the thermostat cover O-ring (arrow) for damage

3 and have it cleaned at a radiator shop.

6   Check the drain bolt gaskets and the O-ring in the thermostat cover (photo) for damage. Replace them with new ones if necessary.

7   Clean the holes, then install the drain bolts and tighten them securely. Apply a thin coat of multi-purpose grease to the thermostat cover O-ring, then check the position of the thermostat and rubber gasket before installing the cover (photo). Tighten the cover screws evenly and securely.

8   Refer to Chapter 4 and remove the fuel tank, then remove the air breather bolt from the T-fitting in the upper radiator hose (photo).

9   Pour a 50/50 mixture of soft water and new ethylene glycol based antifreeze into the radiator until it is full. Reinstall the air breather bolt, the fuel tank and the radiator cap, then start the engine and allow it to run for several minutes — *do not allow it to reach operating temperature.*

10   Stop the engine. Cover the radiator cap with a heavy rag and loosen it to the first stop to allow any pressure in the system to bleed off before the cap is removed completely. Recheck the coolant level in the radiator. If it is low, add more coolant until it reaches the top of the radiator. Reinstall the cap.

11   Add coolant to the reservoir tank until it reaches the Full mark (see Section 5).

12   Start the engine again and allow it to run until it reaches normal operating temperature. Check the system for leaks and recheck the coolant level in the reservoir tank. Install the plastic radiator cover and tighten the screws securely.

13   Do not dispose of the old coolant by pouring it down a drain. Instead, pour it into a heavy plastic container, cap it tightly and take it to an authorized disposal site or a service station.

24   Engine oil/filter — change

1   Consistent routine oil and filter changes are the single most important maintenance procedure you can perform on a motorcycle. The oil not only lubricates the internal parts of the engine, transmission and clutch, but it also acts as a coolant, a cleaner, a sealant, and a protectant. Because of these demands, the oil takes a terrific amount of abuse and should be replaced often with new oil of the recommended grade and type. Saving a little money on the difference in cost between a good oil and a cheap oil won't pay off if the engine is damaged.

2   Before changing the oil and filter, warm up the engine so the oil will drain easily. Be careful when draining the oil, as the exhaust pipes, the engine, and the oil itself can cause severe burns.

3   Put the motorcycle on the centerstand over a clean drain pan. Remove the oil filler cap to vent the crankcase and act as a reminder that there is no oil in the engine.

4   Next, remove the drain plug from the engine (photo) and allow the

23.7 Make sure the thermostat breather and the notch in the gasket (arrow) are positioned exactly as shown before installing the cover

23.8 Cooling system air breather bolt location

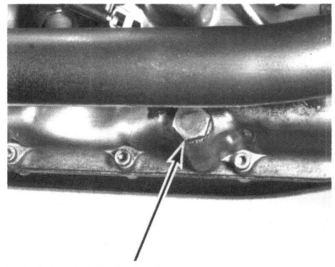

24.4 Engine oil drain plug location

24.5 Oil filter housing screws (arrows)

oil to drain into the pan. Do not lose the sealing washer on the drain plug.

5   As the oil is draining, remove the oil filter housing screws (photo) and lift off the housing and filter. If additional maintenance is planned for this time period, check or service another component while the oil is allowed to drain completely.

6   Clean the filter housing with solvent or clean rags. Wipe any remaining oil off the filter housing sealing area of the crankcase (photo).

7   Check the condition of the drain plug threads and the sealing washer. Use a new O-ring on the filter housing when it is installed (photo).

8   Slide the filter into place on the engine. Coat the O-ring with clean engine oil, attach the filter housing to the engine and tighten the screws evenly and securely.

9   Slip the sealing washer over the drain plug, then install and tighten the plug. If a torque wrench is available, tighten the drain plug to the specified torque. Avoid overtightening, as damage to the engine case will result.

10   Before refilling the engine, check the old oil carefully. If the oil was drained into a clean pan, small pieces of metal or other material can be easily detected. If the oil is very metallic colored, then the engine is experiencing wear from break-in (new engine) or from insufficient lubrication. If there are flakes or chips of metal in the oil, then something is drastically wrong internally and the engine will have to be disassembled for inspection and repair.

24.6 The filter housing O-ring sealing surface of the crankcase (arrow) must be spotlessly clean to prevent leaks

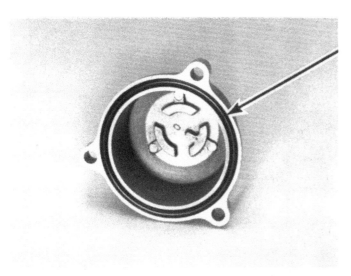

24.7 Make sure the O-ring (arrow) is seated in the housing groove and lubricate it lightly with clean oil

11   If there are pieces of fiber-like material in the oil, the clutch is experiencing excessive wear and should be checked.

12   If the inspection of the oil turns up nothing unusual, refill the crankcase to the proper level with the recommended oil and install the filler cap. Start the engine and let it run for two or three minutes. Shut it off, wait a few minutes, then check the oil level. If necessary, add more oil to bring the level up to the Maximum mark. Check around the drain plug and filter housing for leaks.

13   The old oil drained from the engine cannot be reused in its present state and should be disposed of. Oil reclamation centers, auto repair shops and gas stations will normally accept the oil (which can be refined and used again). After the oil has cooled, it can be drained into a suitable container (capped plastic jugs, topped bottles, milk cartons, etc.) for transport to one of these disposal sites.

25 Lubrication — general

1   Since the controls, cables and various other components of a motorcycle are exposed to the elements, they should be lubricated periodically to ensure safe and trouble-free operation.

2   The footpeg, clutch and brake lever, brake pedal, shift lever and side and centerstand pivots should be lubricated frequently. In order for the lubricant to be applied where it will do the most good, the component should be disassembled. However, if chain and cable lubricant is being used, it can be applied to the pivot joint gaps and will usually work its way into the areas where friction occurs. If motor oil or light grease is being used, apply it sparingly as it may attract dirt (which could cause the controls to bind or wear at an accelerated rate). **Note:** *One of the best lubricants for the control lever pivots is a dry-film lubricant (available from many sources by different names).*

3   The clutch cable should be separated from the handlebar lever and bracket before it is lubricated (photo). It should be treated with motor oil or a commercially available cable lubricant which is specially formulated for use on motorcycle control cables. Small adapters for pressure lubricating the cables with spray can lubricants are available and ensure that the cable is lubricated along its entire length (photo). If motor oil is being used, tape a funnel-shaped piece of heavy paper or plastic to the end of the cable, then pour oil into the funnel and suspend the end of the cable upright. Leave it until the oil runs down into the cable and out the other end. When attaching the cable to the lever, be sure to lubricate the barrel-shaped fitting at the end.

4   To lubricate the throttle cable, disassemble the throttle housing at the handlebar (it is held in place with two screws). Slip the cable end out of the throttle grip and suspend the forward part of the housing and cable from the handlebar. Fill the cable well with lubricant (photo) and let it sit until the oil runs down into the cable. Before reassembling the throttle housing, lubricate the handlebar with oil or lightweight grease. Tighten the mounting screws so the gap between the two parts

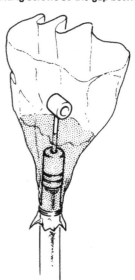

Fig. 1.3   Lubricating a control cable with a makeshift funnel and motor oil (Sec 25)

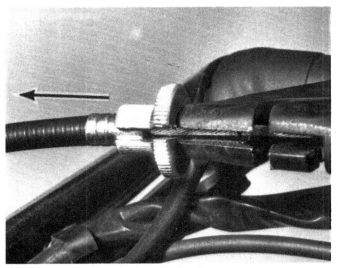

25.3a To remove the clutch cable, line up the slots in the bracket, lock wheel and adjuster, then pull the cable in the direction of the arrow and slide it through the slots

25.3b Lubricating the clutch cable with a pressure luber (make sure the tool seals around the inner cable)

26.3 Removing the air filter housing mounting screws (one at each corner)

of the housing is equal at the top and bottom.

5   Speedometer and tachometer cables should be removed from their housings and lubricated with motor oil or cable lubricant.

26 Air filter — servicing

1   Make sure the lever on the petcock is in the On or Reserve position, then remove the fuel line from the petcock (have a rag handy to catch any spilled fuel). Detach the vacuum hose from the petcock.

2   Remove the side covers and raise the seat, then remove the fuel tank mounting bolt. Slide the tank to the rear and off the motorcycle (be careful not to scratch or otherwise damage the tank).

3   Remove the air filter housing strap and the four cover screws (photo), then lift off the cover. The air filter element can now be lifted out.

4   Tap the filter element on a solid surface to dislodge dirt and dust from the paper. If compressed air is available, use it to clean the element by blowing from the inside out. If the paper is extremely dirty or torn, replace the element with a new one.

5   Reinstall the filter by reversing the removal procedure. Make sure the element is seated properly on the airbox (photo) before installing the cover (the tabs on the filter element must fit between the plastic posts on the airbox).

25.4 When lubricating the throttle cable, fill the area indicated by the arrow with light motor oil

26.5 When installing the filter, the metal tabs on the filter must fit between the plastic posts (arrows) on each side of the airbox

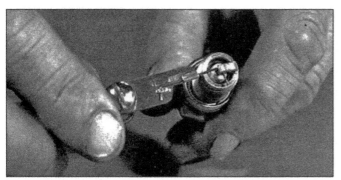

Electrode gap check - use a wire type gauge for best results

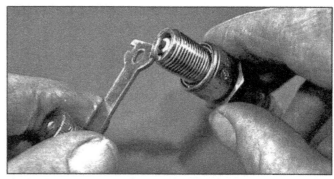

Electrode gap adjustment - bend the side electrode using the correct tool

Normal condition - A brown, tan or grey firing end indicates that the engine is in good condition and that the plug type is correct

Ash deposits - Light brown deposits encrusted on the electrodes and insulator, leading to misfire and hesitation. Caused by excessive amounts of oil in the combustion chamber or poor quality fuel/oil

Carbon fouling - Dry, black sooty deposits leading to misfire and weak spark. Caused by an over-rich fuel/air mixture, faulty choke operation or blocked air filter

Oil fouling - Wet oily deposits leading to misfire and weak spark. Caused by oil leakage past piston rings or valve guides (4-stroke engine), or excess lubricant (2-stroke engine)

Overheating - A blistered white insulator and glazed electrodes. Caused by ignition system fault, incorrect fuel, or cooling system fault

Worn plug - Worn electrodes will cause poor starting in damp or cold weather and will also waste fuel

27 Final drive gear oil — change

1   Ride the motorcycle for several miles at highway speeds before performing this operation. This will ensure that the old oil and contaminants drain completely.
2   Clean the area around the final drive gear housing oil drain plug and filler cap. This will help prevent the entry of dirt when the plugs are removed. Place a clean drain pan under the housing, then remove the filler cap and the drain plug (photo). Allow the oil to drain into the pan, then clean the plug and drain hole thoroughly. Reinstall the plug and tighten it securely.
3   Pour new gear oil of the recommended grade and type into the filler hole until the level reaches the upper outside edge of the hole (refer to Section 7 if necessary).
4   Clean the filler cap, then reinstall and tighten it securely.

28 Cylinder compression — check

1   Among other things, poor engine performance may be caused by leaking valves, incorrect valve clearances, a leaking head gasket, or worn pistons, rings and/or cylinder walls. A cylinder compression check will help pinpoint these conditions and can also indicate the presence of excessive carbon deposits in the cylinder heads.
2   The only tools required are a compression gauge and a spark plug wrench. Depending on the outcome of the initial test, a squirt-type oil can may also be needed.
3   Start the engine and allow it to reach normal operating temperature. Place the motorcycle on the centerstand and remove both spark plugs. Work carefully, do not strip the spark plug hole threads and do not burn your hands. Use jumper wires to ground the plug wires to the engine.
4   Install the compression gauge in one of the spark plug holes (photo). Hold or block the throttle wide open.
5   Crank the engine over a minimum of four or five revolutions (or until the gauge reading stops increasing) and observe the intial movement of the compression gauge needle as well as the final total gauge reading. Repeat the procedure for the remaining cylinder and compare the results to the Specifications.
6   If the compression in both cylinders built up quickly and evenly to the specified amount, you can assume that the engine upper end is in reasonably good mechanical condition. Worn or sticking piston rings and worn cylinders will produce very little initial movement of the gauge needle, but compression will tend to build up gradually as the engine spins over. Valve and valve seat leakage, or head gasket leakage, is indicated by low initial compression which does not tend to build up.
7   To further confirm your findings, add a small amount of engine oil to each cylinder by inserting the nozzle of a squirt-type oil can through

27.2 Final drive gear housing oil drain plug location

28.4 Access to the spark plug holes is restricted, so work carefully when installing the compression gauge or gauge adapter (arrow) to avoid stripping the threads

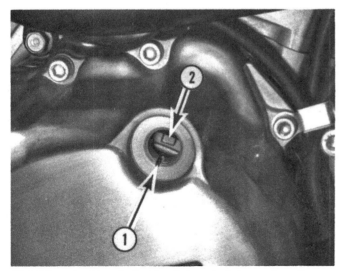

29.4 When checking the ignition timing, the index mark on the crankcase (1) must fall within the firing range mark on the rotor (2)

30.2 Make sure the wrench or socket fits properly before removing the plugs

the spark plug holes. The oil will tend to seal the piston rings if they are leaking. Repeat the test for both cylinders.

8   If the compression increases significantly after the addition of the oil, the piston rings and/or cylinders are definitely worn. If the compression does not increase, the pressure is leaking past the valves or the head gasket. Leakage past the valves may be due to insufficient valve clearances, burned, warped or cracked valves or valve seats or valves that are hanging up in the guides.

9   If compression readings are considerably higher than specified, the combustion chambers are probably coated with excessive carbon deposits. It is possible for carbon deposits to raise the compression enough to compensate for the effects of leakage past rings or valves. Refer to Chapter 2, remove the cylinder heads and carefully decarbonize the combustion chambers.

29 Ignition timing — check

1   The ignition timing is factory preset and cannot be adjusted. If the timing is not correct, further inspection of the TCI system should be carried out to determine which component has failed. Refer to Chapter 5.

2   The timing must be checked using a strobe type timing light (preferably with an inductive pickup) and the engine must be running,

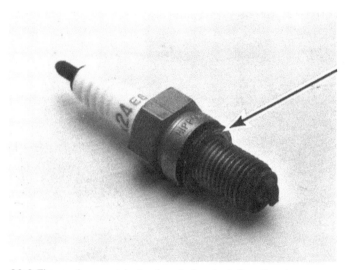

30.6 The washer must be in place before installing the plug

31.2 Push the left side of the radiator forward and hold it in position with an elastic cord, wire or rope

so it is best done is a well-ventilated area (try to stay away from direct sunlight).

3   Make sure the engine has been warmed up to normal operating temperature and check the idle speed to see if it is correct. Place the motorcycle on the centerstand and remove the timing plug and the rectangular cover from the left engine side cover.

4   Place a 14 mm socket and ratchet over the alternator rotor bolt and slowly turn the rotor counterclockwise until the firing range mark (photo) appears in the window. Carefully fill in the indented lines with white paint so they will be easily visible when the timing light flashes.

5   Hook up the timing light according to the manufacturer's instructions. Generally the red lead should be hooked to the positive post of the battery, the black lead should be hooked to the negative post of the battery and the remaining lead to the number one (front cylinder) spark plug lead. Since the battery posts on the motorcycle are relatively inaccessible, you may want to use a separate 12-volt battery (such as a car battery) to power the timing light.

6   Start the engine and direct the timing light through the window at the rotor.

7   With the engine idling, the index mark on the crankcase should be within the firing range mark on the rotor. If it is not, the ignition timing is incorrect. Double check the idle speed and make sure it is as specified.

8   If the ignition timing is incorrect, check the TCI unit and the alternator components as described in Chapter 5. Replace any defective parts with new ones.

30 Spark plugs — replacement

1   This motorcycle is equipped with spark plugs that have 12 mm threads and an 18 mm wrench hex. Make sure that your spark plug socket is the correct size before attempting to remove the plugs.

2   Disconnect the spark plug caps and remove the plugs (photo).

3   Inspect the electrodes for wear. Both the center and side electrodes should have square edges and the side electrode should be of uniform thickness. Look for excessive deposits and evidence of a cracked or chipped insulator around the center electrode. Compare your spark plugs to the   spark plug reading chart. Check the threads, the washer and the procelain insulator body for cracks and other damage.

4   If the electrodes are not excessively worn, and if the deposits can be easily removed with a wire brush, the plugs can be regapped and reused (if no cracks or chips are visible in the insulator). If in doubt concerning the condition of the plugs, replace them with new ones, as the expense is minimal.

5   Cleaning spark plugs by sandblasting is not recommended, as grit from the sandblasting process may remain in the plug and be dislodged after the plug is installed in the engine, which obviously can cause damage and increased wear.

6   Before installing new plugs, make sure they are the correct type and heat range. Check the gap between the electrodes, as they are not preset. For best results, use a wire-type gauge rather than a flat gauge to check the gap. If the gap must be adjusted, bend the side electrode only and be very careful not to chip or crack the insulator nose. Make sure the washer (photo) is in place before installing each plug.

7   Since the cylinder heads are made of aluminum, which is soft and easily damaged, thread the plugs into the heads by hand. Once the plugs are finger tight, the job can be finished with a wrench. If a torque wrench is available, tighten the spark plugs to the specified torque. If you do not have a torque wrench, tighten the plugs finger tight (until the washers bottom on the cylinder head) then use a wrench to tighten them an additional 1/4 turn. Regardless of the method used, do not over tighten them.

8   Reconnect the spark plug caps.

31 Valve clearances — check and adjustment

Note: *This motorcycle engine is equipped with a valvetrain which requires changing shims located on top of the lifters to vary the valve clearances. As a result, special tools, shims and a rather complicated procedure are involved. If you do not have access to the necessary tools or if you do not desire to tackle the procedure, take the machine to a reputable Yamaha dealer service department for the valve clearance adjustment. If you plan on doing the job yourself, read through the en-*

tire procedure before starting work. This will enable you to plan ahead, anticipate what is coming and avoid problems.

1   The engine must be completely cool for this maintenance procedure, so let the machine sit overnight before beginning. Refer to Chapter 4 and remove the fuel tank. Remove the rubber strap from the airbox, then separate the front crankcase vent hose from the cylinder head cover fitting. Separate the rear vent hose from the airbox, then loosen the clamp band screws that attach the airbox to the carburetors and lift off the airbox. Cover the carburetor openings with a clean rag to keep foreign objects out.

2   Remove the plastic radiator cover (it is held in place with four screws) and the radiator mounting bolts. It is not necessary to remove the radiator from the motorcycle, just push the left side gently but firmly toward the front wheel (photo). The right side of the radiator and the connected hoses should remain in place.

3   Remove the plastic air baffle plate and the right side frame tube. Disconnect both spark plug wires and remove the plugs. Remove both cylinder head covers (they are held in place with four bolts each). Do not lose the camshaft oil plugs (photo) when the covers are lifted off the cylinder heads.

4   Remove the rectangular cover and the timing plug from the left engine side cover. Slip a 14 mm socket and ratchet over the rotor bolt and turn the crankshaft counterclockwise very slowly until the T index mark is aligned with the crankcase mark in the window (photo).

Next, check the cam lobes for the front cylinder. If they are pointing up (photo) (no pressure on the lifters), then the front piston is at top dead center (TDC) on the compression stroke and the valve clearances can be checked (front cylinder only). If the cam lobes are not pointing up, turn the crankshaft counterclockwise one complete revolution (360°) and align the marks as described previously.

5   Slip feeler gauges of various thicknesses between each camshaft lobe and lifter (photo) to determine the clearances (since the accompanying shim selection chart is in millimeters, the use of metric size feeler gauges will make the procedure easier). The feeler gauge which denotes the clearance should just slide between the cam and shim with a slight drag. Record the measured clearance for each valve on the appropriate *Measured clearance* line of the accompanying chart.

6   Compare the measured clearances to the Specifications at the front of this Chapter (note that intake and exhaust valves require different clearances). If the clearances are within the specified range, no action is required. If any of the clearances are outside the specified range (which is probably the case), the shims must be replaced with thicker or thinner ones to obtain the correct clearances. The next step is to determine and record the shim sizes in the front cylinder. Once that is done, the valve clearances and shim sizes for the rear cylinder must be obtained. Then only one trip to a Yamaha dealer will be needed to buy any necessary shims and complete the valve adjustment procedure.

7   Rotate the lifters in the front head until the slots are opposite each

31.3 Do not lose the oil plugs (arrow) when the cylinder head covers are removed (they must be in place when the covers are reinstalled)

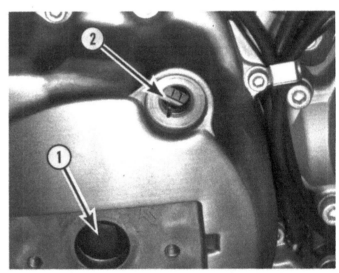

31.4a Turn the bolt (1) counterclockwise to align the T index mark with the crankcase index mark (2)

31.4b If the cam lobes (arrows) are pointing up as shown, the valve clearances can be checked

31.5 Checking the clearance of one of the exhaust valves with a feeler gauge

FRONT

NO. 1 CYL.

Measured clearance _____

Installed shim _____

Required shim _____

New clearance _____

Measured clearance _____

Installed shim _____

Required shim _____

New clearance _____

EXHAUST VALVES

Measured clearance _____

Installed shim _____

Required shim _____

New clearance _____

Measured clearance _____

Installed shim _____

Required shim _____

New clearance _____

INTAKE VALVES

NO. 2 CYL.

Measured clearance _____

Installed shim _____

Required shim _____

New clearance _____

Measured clearance _____

Installed shim _____

Required shim _____

New clearance _____

INTAKE VALVES

Measured clearance _____

Installed shim _____

Required shim _____

New clearance _____

Measured clearance _____

Installed shim _____

Required shim _____

New clearance _____

EXHAUST VALVES

REAR

Fig. 1.4 DO NOT WRITE ON THIS CHART — make a photocopy of this page and record the valve adjustment data on the copy (then the chart can be used more than once) (Sec 31)

other and facing the center of the cylinder head (photo). Locate the camshaft identification marks (I for intake/E for exhaust), which should be facing up. Turn the crankshaft slowly in a clockwise direction until the identical mark stamped into the opposite side of the camshaft, 180° from the first mark, is visible (photo).

8   Position the lifter depressing tool over the exhaust camshaft with the match mark (circular indentation) on the tool (photo) adjacent to the (now visible) lower camshaft identification mark. The tool ridge must slide between the lifters and the tool lobes must rest against the tops of the lifters. **Caution:** *Make sure the tool lobes do not ride up on the shims (photo).*

9   Turn the crankshaft very slowly counterclockwise and the tool will depress the lifters slightly. Make sure that the tool does not gouge the cover gasket sealing surface of the cylinder head.

10   The shims can now be removed from the exhaust valve lifters. Slip a scriber or very small screwdriver through the slot and under one of the shims (photo), then lift it out with a forceps.

11   The shim size is etched into the back side of the shim (photo); record it on the appropriate *Installed shim* line of the accompanying chart. If the number cannot be read, measure the shim thickness with a 0-to-1 inch micrometer. Place the shim back into the lifter with the number *down*, then remove the remaining exhaust valve shim and record its size. **Caution:** *Do not turn the crankshaft without shims in place in the lifters — damage to the camshaft and lifters will result if you do.*

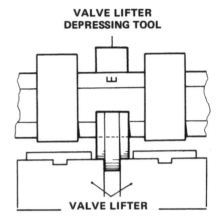

Fig. 1.5   The valve lifter depressing tool must contact the lifters exactly as shown (Sec 31)

31.7a Rotate the lifters until the slots face the center of the head

31.7b As the crankshaft is turned clockwise the opposite EX mark will come into view

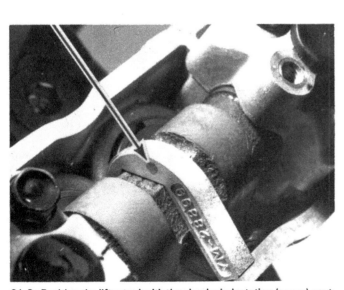

31.8a Position the lifter tool with the circular indentation (arrow) next to the EX mark on the camshaft . . .

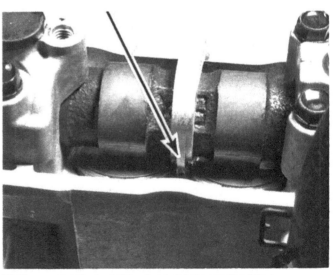

31.8b . . . and the tool ridge (arrow) between the lifters

Exhaust

MEASURED CLEARANCE	200	205	210	215	220	225	230	235	240	245	250	255	260	265	270	275	280	285	290	295	300	305	310	315	320
INSTALLED SHIM NUMBER*																									
0.00 ~ 0.05				200	205	210	215	220	225	230	235	240	245	250	255	260	265	270	275	280	285	290	295	300	305
0.06 ~ 0.10			200	205	210	215	220	225	230	235	240	245	250	255	260	265	270	275	280	285	290	295	300	305	310
0.11 ~ 0.15		200	205	210	215	220	225	230	235	240	245	250	255	260	265	270	275	280	285	290	295	300	305	310	315
0.16 ~ 0.20																									
0.21 ~ 0.25	205	210	215	220	225	230	235	240	245	250	255	260	265	270	275	280	285	290	295	300	305	310	315	320	
0.26 ~ 0.30	210	215	220	225	230	235	240	245	250	255	260	265	270	275	280	285	290	295	300	305	310	315	320		
0.31 ~ 0.35	215	220	225	230	235	240	245	250	255	260	265	270	275	280	285	290	295	300	305	310	315	320			
0.36 ~ 0.40	220	225	230	235	240	245	250	255	260	265	270	275	280	285	290	295	300	305	310	315	320				
0.41 ~ 0.45	225	230	235	240	245	250	255	260	265	270	275	280	285	290	295	300	305	310	315	320					
0.46 ~ 0.50	230	235	240	245	250	255	260	265	270	275	280	285	290	295	300	305	310	315	320						
0.51 ~ 0.55	235	240	245	250	255	260	265	270	275	280	285	290	295	300	305	310	315	320							
0.56 ~ 0.60	240	245	250	255	260	265	270	275	280	285	290	295	300	305	310	315	320								
0.61 ~ 0.65	245	250	255	260	265	270	275	280	285	290	295	300	305	310	315	320									
0.66 ~ 0.70	250	255	260	265	270	275	280	285	290	295	300	305	310	315	320										
0.71 ~ 0.75	255	260	265	270	275	280	285	290	295	300	305	310	315	320											
0.76 ~ 0.80	260	265	270	275	280	285	290	295	300	305	310	315	320												
0.81 ~ 0.85	265	270	275	280	285	290	295	300	305	310	315	320													
0.86 ~ 0.90	270	275	280	285	290	295	300	305	310	315	320														
0.91 ~ 0.95	275	280	285	290	295	300	305	310	315	320															
0.96 ~ 1.00	280	285	290	295	300	305	310	315	320																
1.01 ~ 1.05	285	290	295	300	305	310	315	320																	
1.06 ~ 1.10	290	295	300	305	310	315	320																		
1.11 ~ 1.15	295	300	305	310	315	320																			
1.16 ~ 1.20	300	305	310	315	320																				
1.21 ~ 1.25	305	310	315	320																					
1.26 ~ 1.30	310	315	320																						
1.31 ~ 1.35	315	320																							
1.36 ~ 1.40	320																								

VALVE CLEARANCE (engine cold)
 0.16 ~ 0.20 mm
Example: Installed is 250
 Measured clearance is 0.32 mm
 Replace 250 with 265
*Shim number (example):
 Pad No. 250 = 2.50 mm
 Pad No. 255 = 2.55 mm
Always install shim with number down.

Fig. 1.6 Exhaust valve shim selection chart (Sec 31)

Intake

MEASURED CLEARANCE	200	205	210	215	220	225	230	235	240	245	250	255	260	265	270	275	280	285	290	295	300	305	310	315	320
INSTALLED SHIM NUMBER*																									
0.00 ~ 0.05			200	205	210	215	220	225	230	235	240	245	250	255	260	265	270	275	280	285	290	295	300	305	310
0.06 ~ 0.10		200	205	210	215	220	225	230	235	240	245	250	255	260	265	270	275	280	285	290	295	300	305	310	315
0.11 ~ 0.15																									
0.16 ~ 0.20	205	210	215	220	225	230	235	240	245	250	255	260	265	270	275	280	285	290	295	300	305	310	315	320	
0.21 ~ 0.25	210	215	220	225	230	235	240	245	250	255	260	265	270	275	280	285	290	295	300	305	310	315	320		
0.26 ~ 0.30	215	220	225	230	235	240	245	250	255	260	265	270	275	280	285	290	295	300	305	310	315	320			
0.31 ~ 0.35	220	225	230	235	240	245	250	255	260	265	270	275	280	285	290	295	300	305	310	315	320				
0.36 ~ 0.40	225	230	235	240	245	250	255	260	265	270	275	280	285	290	295	300	305	310	315	320					
0.41 ~ 0.45	230	235	240	245	250	255	260	265	270	275	280	285	290	295	300	305	310	315	320						
0.46 ~ 0.50	235	240	245	250	255	260	265	270	275	280	285	290	295	300	305	310	315	320							
0.51 ~ 0.55	240	245	250	255	260	265	270	275	280	285	290	295	300	305	310	315	320								
0.56 ~ 0.60	245	250	255	260	265	270	275	280	285	290	295	300	305	310	315	320									
0.61 ~ 0.65	250	255	260	265	270	275	280	285	290	295	300	305	310	315	320										
0.66 ~ 0.70	255	260	265	270	275	280	285	290	295	300	305	310	315	320											
0.71 ~ 0.75	260	265	270	275	280	285	290	295	300	305	310	315	320												
0.76 ~ 0.80	265	270	275	280	285	290	295	300	305	310	315	320													
0.81 ~ 0.85	270	275	280	285	290	295	300	305	310	315	320														
0.86 ~ 0.90	275	280	285	290	295	300	305	310	315	320															
0.91 ~ 0.95	280	285	290	295	300	305	310	315	320																
0.96 ~ 1.00	285	290	295	300	305	310	315	320																	
1.01 ~ 1.05	290	295	300	305	310	315	320																		
1.06 ~ 1.10	295	300	305	310	315	320																			
1.11 ~ 1.15	300	305	310	315	320																				
1.16 ~ 1.20	305	310	315	320																					
1.21 ~ 1.25	310	315	320																						
1.26 ~ 1.30	315	320																							
1.31 ~ 1.35	320																								

VALVE CLEARANCE (engine cold)
 0.11 ~ 0.15 mm
Example: Installed is 250
 Measured clearance is 0.32 mm
 Replace 250 with 270
*Shim number (example):
 Pad No. 250 = 2.50 mm
 Pad No. 255 = 2.55 mm
Always install shim with number down.

Fig. 1.7 Intake valve shim selection chart (Sec 31)

12   Turn the crankshaft clockwise slightly to remove the lifter depressing tool from the exhaust camshaft.
13   Repeat Steps 8 through 11 to determine the shim sizes for the front cylinder's intake valves (photo). Be sure to record the sizes on the accompanying chart, then reinstall the shims.
14   Rotate the crankshaft approximately 290° counterclockwise and align the 2T index mark with the crankcase mark in the window (photo). Check the cam lobes for the rear cylinder. If they are pointing up (no pressure on the lifters), then the rear piston is at top dead center (TDC) on the compression stroke and the valve clearances for the rear cylinder can be checked.
15   Repeat Steps 5 through 13 to check and record the valve clearances and shim sizes for the rear cylinder's valves.
16   At this point you should have all of the *Measured clearances* and *Installed shim* blanks on the accompanying chart filled in. Next, put an X through the boxes on the chart which correspond to the valves which *do not* require adjustment.
17   Refer to the accompanying shim selection charts (one for intake valves, one for exhaust valves) and determine what shims are needed to bring the remaining valve clearances into the specified range. Find the measured clearance for a given valve in the vertical column at the left side of the chart and the installed shim size for the same valve in the horizontal column at the top of the chart. Read across from the clearance and down from the shim size until the columns intersect.

The box at that point contains the size of the shim which must be installed in that particular valve lifter to produce the desired clearance. Record the number on the appropriate *Required shim* line of the chart, then repeat the selection process for the remaining valves that require adjustment. Be sure to use the correct chart; *Exhaust* for exhaust valves, *Intake* for intake valves.
18   Once the required shim sizes have been recorded, examine the installed shim sizes entered on the chart to see if some of the shims can be swapped around between the valves. If the clearances cannot be adjusted by swapping shims, then new shims will have to be purchased from your Yamaha dealer. **Caution:** *Do not try to grind shims down to make them usable.*
19   After the required shims have been obtained, install them and *recheck the clearances* following the procedure described previously. Keep in mind that the shims must be installed with the numbers down and never turn the crankshaft unless all of the shims are in place. **Note:** *Before rechecking the clearances, turn the crankshaft through several revolutions to seat the shims in the lifters.* Record the clearances after they have been adjusted, then you will be able to see how rapidly and to what extent they are changing. You can then lengthen or shorten future maintenance intervals accordingly.
20   Make sure the camshaft oil plugs are in place and check the gaskets for damage, then install the cylinder head covers. Note that the cover with the large crankcase vent hose fitting must be installed

31.10 Dislodging a front cylinder exhaust valve shim with a screwdriver

31.11 Shim size number (this one is 2.55 mm thick)

31.13 Removing a front cylinder intake valve shim

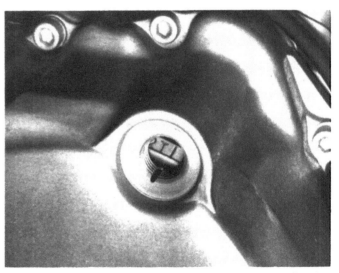
31.14 Before checking the rear cylinder valve clearances, align the 2T index mark with the crankcase index mark

on the rear cylinder head. Tighten the bolts evenly and securely, following a criss-cross pattern.

21   Install the remaining components and hoses by reversing the removal procedure. Be sure to tighten the right side frame tube bolts to the specified torque. Start and run the engine and check for oil leaks around the head cover gaskets.

32   Idle speed — adjustment

1   The idle speed should be checked and adjusted after the carburetors are synchronized and when it is obviously too high or too low. Before adjusting the idle speed, make sure the valve clearances and spark plug gaps are correct and check the ignition timing.

2   The engine should be at normal operating temperature, which is usually reached after 10 to 15 minutes of stop and go riding. Place the motorcycle on the centerstand and make sure the transmission is in Neutral.

3   Turn the throttle stop screw on the rear carburetor (photo), until the specified idle speed is obtained.

4   If a smooth, steady idle cannot be achieved, the fuel/air mixture may be incorrect. Refer to Chapter 4 for additional carburetor information.

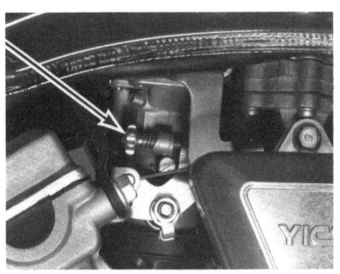

32.3 Throttle stop screw location (rear carburetor)

33.9   Remove the plug (1) from the rear intake manifold and the vacuum hose (2) from the front intake manifold to connect the manometer; to adjust, loosen the locknut (3) and turn the rod (4)

33   Carburetors — synchronization

1   Carburetor synchronization is simply the process of adjusting the carburetors so they pass the same amount of fuel/air mixture to each cylinder. This is done by measuring the vacuum produced in each cylinder. Carburetors that are out of synchronization will result in decreased fuel mileage, increased engine temperature, less than ideal throttle response and higher vibration levels.

2   To properly synchronize the carburetors, you will need some sort of vacuum gauge setup, preferably with a gauge for each cylinder, or a mercury manometer, which is a calibrated tube arrangement that utilizes columns of mercury to indicate engine vacuum.

3   A manometer can be purchased from a motorcycle dealer or accessory shop and should have the necessary rubber hoses supplied with it for hooking into the intake manifolds of the engine.

4   A vacuum gauge setup can also be purchased from a dealer or fabricated from commonly available hardware and automotive vacuum gauges.

5   The manometer is the more reliable and accurate instrument, and for that reason is preferred over the vacuum gauge setup; however, since the mercury used in the manometer is a liquid, and extremely toxic, extra precautions must be taken during use and storage of the instrument.

6   Because of the nature of the synchronization procedure and the need for special instruments, most owners leave the task to a dealer service department or a reputable motorcycle repair shop.

7   Place the fuel petcock lever in the Prime position.

8   Start the engine and let it run until it reaches normal operating temperature, then shut it off.

9   Remove the plug (photo) from the intake manifold fittings, then hook up the vacuum gauge set or the manometer according to the manufacturer's instructions. Make sure there are no leaks in the setup, as false readings will result.

10   Start the engine and make sure the idle speed is correct.

11   The vacuum readings for both of the cylinders should be the same (photo). If the vacuum readings vary, adjust as necessary.

12   To perform the adjustment, loosen the locknut on the synchronizing rod (photo 33.9) and turn the rod as necessary, until the vacuum is identical or nearly identical for each cylinder. The rod can be turned by attaching a very small adjustable wrench to the flats milled into the center of the rod.

13   When the adjustment is complete, retighten the synchronizing rod locknut, recheck the vacuum reading and idle speeds, then stop the engine. Remove the vacuum gauge hoses and attach the plug and hose to the intake manifolds. Turn the fuel petcock lever to the On position.

34   Wheel bearings — check and lubrication

1   The wheel bearings should be checked at the recommended inter-

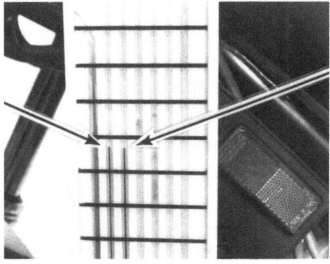

33.11 The mercury columns in the manometer (arrows) should be the same height (if not, adjustment is required)

35.2a Loosen all eight (four on each fork leg) pinch bolts before removing the fork caps

35.2b A 17 mm nylon locknut can be used to remove the fork cap bolts (RJ models)

vals or whenever a rumbling noise, which increases with increased wheel speed, is noticed.

2   To check the front wheel bearings, place the motorcycle on the centerstand and have an assistant sit on the rear of the seat to raise the front wheel off the ground.

3   Spin the wheel by hand and touch the axle or front fender lightly. If a rumbling type vibration is felt, the wheel should be removed and the bearings checked and replaced (see Chapter 7).

4   To check the rear wheel bearings the wheel must be removed from the machine (refer to Chapter 7). If radial or axial play can be felt as the bearings are turned with your fingers, they should be replaced with new ones (again, refer to Chapter 7).

5   Lubrication of the bearings (or repacking, as it is sometimes known) does not have to be done very often. It is a rather involved procedure requiring removal of the wheels and careful removal and installation of the bearings (refer to Chapter 7 for the procedure).

35   Fork oil – replacement

Note: *On RK models, bleed off all fork air pressure by depressing the air valve with a small screwdriver before removing the fork cap bolts.*

1   Place the motorcycle on the centerstand and position a jack or blocks of wood under the front of the engine to support the motorcycle when the fork cap bolts are removed. Remove the wire holders from the upper fork tube pinch bolts.

2   Loosen the upper pinch bolts (photo) and remove the rubber caps (RJ models) and cap bolts from the fork tubes. The upper pinch bolts are very tight and will require an impact driver or a 1/2 inch drive Allen head socket to loosen them. On RJ models, a 17 mm Allen wrench is needed to remove the cap bolts but a 10 mm nut (which is 17 mm across the flats) can be used as a substitute tool (photo). Place a smaller nut in the bottom of the cap before inserting the 10 mm nut. The small nut will act as a spacer and hold the large nut part way out of the cap so it can be turned with a socket. Push down on the ratchet as the last couple of threads are disengaged, as the cap is under slight spring pressure.

3   Place a drain pan under each fork leg and remove the drain screws (photo). **Warning:** *Do not allow the fork oil to contact the brake disc or pads. If it does, clean the disc with alcohol or lacquer thinner and replace the pads with new ones before riding the motorcycle.*

4   After most of the oil has drained, slowly compress and release

35.3 Fork oil drain screw location

the forks to pump out the remaining oil. An assistant will most likely be required to do this procedure.

5   Check the drain screw gaskets for damage and replace them if necessary. Reinstall the screws in the fork legs and tighten them securely.

6   Pour the specified amount of oil into each of the fork tubes through the openings at the top. **Note:** *For optimum fork performance, remove the springs before adding new fork oil, then compress the forks completely and add oil until the level is the specified distance from the top of the fork tube – it should be exactly the same in each fork leg. Be sure to use oil that is formulated for use in front suspensions. Slowly pump the forks up and down to distribute the oil.*

7   Check the O-rings on the fork cap bolts, then coat them with a thin layer of multi-purpose grease. Install the bolts and tighten them securely. Apply pressure to the cap bolts as the first few threads are engaged in the fork tubes. Push the rubber caps into place in the bolts (RJ models).

8   Tighten the fork tube pinch bolts to the specified torque and install the wire holders. **Note:** *Tighten the front bolts first, then the rear bolts; the gap between the handlebar forging and the upper triple clamp must be at the rear (see Chapter 6).*

Chapter 2 Engine and transmission

Contents

Specifications

General

Bore .... 3.150 in (80.0 mm)
Stroke.. 2.165 in (55.0 mm)
Displacement .... 552 cc
Compression ratio .... 10.5 : 1

Camshaft

Bearing journal diameter .... 0.9830 to 0.9835 in (24.960 to 24.980 mm)
Bearing oil clearance
    Standard .... 0.0008 to 0.0024 in (0.020 to 0.061 mm)
    Service limit .... 0.006 in (0.160 mm)
Lobe height
    Standard
        Intake .... 1.449 ± 0.002 in (36.80 ± 0.05 mm)
        Exhaust .... 1.429 ± 0.002 in (36.30 ± 0.05 mm)
    Service limit
        Intake .... 1.443 in (36.65 mm)
        Exhaust .... 1.423 in (36.15 mm)
Runout limit .... 0.0012 in (0.03 mm)

Valves, valve springs and cylinder heads

Valve face width .... 0.090 in (2.3 mm)
Valve seat width
    Standard .... 0.040 ± 0.004 in (1 ± 0.1 mm)
    Service limit .... 0.067 in (1.7 mm)
Margin thickness limit .... 0.040 ± 0.008 in (1 ± 0.2 mm) min.
Valve stem diameter
    Intake .... 0.2390 to 0.2396 in (5.975 to 5.990 mm)
    Exhaust .... 0.2384 to 0.2390 in (5.960 to 5.975 mm)
Valve guide diameter (intake and exhaust) .... 0.240 to 0.2405 in (6.0 to 6.01 mm)
Valve stem-to-guide clearance
    Intake .... 0.0004 to 0.0015 in (0.010 to 0.037 mm)
    Exhaust .... 0.001 to 0.002 in (0.025 to 0.052 mm)
Valve stem runout limit .... 0.0004 in (0.01 mm)
Valve spring free length (intake and exhaust)
    Inner spring .... 1.424 in (36.17 mm)
    Outer spring .... 1.442 in (36.63 mm)
Valve spring out-of-square limit .... 2.5°/0.063 in (1.6 mm)

Cylinders

Bore size
    Standard .... 3.150 in (80 mm)
    Service limit .... 3.154 in (80.1 mm)
Taper limit .... 0.0002 in (0.005 mm)
Out-of-round limit .... 0.0004 in (0.01 mm)
Cylinder-to-piston clearance .... 0.0020 to 0.0028 in (0.050 to 0.070 mm)

Pistons and piston rings

Piston diameter .... 3.1476 to 3.1484 in (79.94 to 79.96 mm)
Piston ring-to-groove clearance
    Top ring .... 0.0012 to 0.0028 in (0.03 to 0.07 mm)
    2nd ring .... 0.0008 to 0.0024 in (0.02 to 0.06 mm)
    Oil ring .... No measurable clearance
Piston ring end gap
    Top/2nd rings .... 0.012 to 0.020 in (0.3 to 0.5 mm)
    Oil ring side rails .... 0.012 to 0.035 in (0.3 to 0.9 mm)

Crankshaft and bearings

Connecting rod side clearance .... 0.013 to 0.017 in (0.32 to 0.43 mm)
Crankshaft runout limit .... 0.0008 in (0.002 mm)
Main bearing oil clearance .... 0.0008 to 0.0024 in (0.020 to 0.062 mm)
Connecting rod bearing oil clearance .... 0.0014 to 0.0021 in (0.035 to 0.054 mm)

Clutch

Friction plate thickness
    Standard .... 0.110 ± 0.003 in (2.8 ±0.08 mm)
    Service limit .... 0.100 in (2.6 mm)
Friction plate quantity 8
Metal plate thickness .... 0.060 in (1.6 mm)
Metal plate warpage limit .... 0.008 in (0.2 mm)
Metal plate quantity 7
Clutch spring free length
    Standard .... 1.620 in (41.2 mm)
    Service limit .... 1.583 in (40.2 mm)
Pushrod runout limit .... 0.020 in (0.5 mm)

Oil pump

Inner rotor-to-outer rotor clearance .... 0.0047 in (0.12 mm) max.
Outer rotor-to-pump body clearance .... 0.0012 to 0.0035 in (0.03 to 0.09 mm) max.

Torque specifications

	Ft-lb	Nm
Cylinder head nut	36.2	50
Cylinder head bolt	7.2	10
Camshaft bearing cap bolt	5.8	8
Camshaft sprocket bolt	50.0	70
Cam chain tensioner mounting bolts	8.7	12
Cam chain tensioner end plug	14.0	20
Cylinder head cover bolt	7.2	10
Connecting rod nuts	27.0	38
Balancer drive gear (left side of crankshaft)	72.0	100
Balancer shaft nut	43.0	60
Starter clutch bolts	14.0	20
Alternator rotor bolt	58.0	80

Torque specifications (continued)

	Ft-lb	Nm
Cam chain tensioner slipper bolts	5.8	8
Timing gear shaft bolt	8.7	12
Oil line banjo fitting union bolt	13.0	18
Oil pump bolts	7.2	10
Primary drive gear nut (right end of crankshaft)	50.0	70
Clutch center hub nut	50.0	70
Clutch release plate bolts	5.8	8
Crankcase bolts		
10 mm	22.0	30
6 mm	7.2	10
Intake manifold bolts	8.7	12
Rear cylinder upper engine mount bracket bolts	14.0	20
Frame tube-to-frame bolts	19.0	26
Upper rear bracket-to-frame bolts	14.0	20
Engine-to-frame (center)	42.0	58
Engine-to-frame (rear)	35.0	48
Sidestand bracket-to-engine	40.0	55
Footpeg/engine mount stud nuts	50.0	70
Muffler/passenger footpeg-to-bracket bolts	31.0	42
Exhaust pipe-to-cylinder head flange bolts	7.2	10
Muffler clamp bolt	14	20

1   General information

The Yamaha Vision is powered by a twin cylinder, liquid-cooled, DOHC engine with the cylinders arranged in a 70° V-configuration. The cylinder heads are equipped with four valves per cylinder.

A counterbalancer, which cancels the vibration produced by this type engine, is mounted in the front of the crankcase and driven by gears from the left end of the crankshaft.

The crankshaft is mounted in 360° insert-type plain bearings and the connecting rods ride in split-type plain bearings on a common crankshaft journal.

The camshafts for each cylinder ride directly in the cylinder head material and are chain driven. Tensioning of the chains is done automatically, with no adjustment or maintenance required. The cam lobes bear directly on shims mounted in the valve lifters. Valve adjustment is accomplished by replacing the shims.

The crankcases are die-cast from aluminum and split vertically for access to the lower end and transmission components. The transmission is a five-speed, constant mesh unit coupled to shaft final drive through a bevel gear output gearbox. **Note:** *If service or repair of the output gearbox is required, it must be done by a Yamaha dealer service department.*

2   Major engine repair — general note

1   It is not always easy to determine when or if an engine should be completely overhauled, as a number of factors must be considered.
2   High mileage is not necessarily an indication that an overhaul is needed, while low mileage, on the other hand, does not preclude the need for an overhaul. Frequency of servicing is probably the single most important consideration. An engine that has regular and frequent oil and filter changes, as well as other required maintenance, will most likely give many miles of reliable service. Conversely, a neglected engine, or one which has not been broken in properly, may require an overhaul very early in its life.
3   Exhaust smoke and excessive oil consumption are both indications that piston rings and/or valve guides are in need of attention. Make sure that oil leaks are not responsible before deciding that the rings and guides are bad. Refer to Chapter 1 and perform a cylinder compression check to determine for certain the nature and extent of the work required.
4   If the engine is making obvious knocking or rumbling noises, the connecting rod and/or main bearings are probably at fault.
5   Loss of power, rough running, excessive valve train noise and high fuel consumption rates may also point to the need for an overhaul, especially if they are all present at the same time. If a complete tune-up does not remedy the situation, major mechanical work is the only solution.
6   An engine overhaul generally involves restoring the internal parts to the specifications of a new engine. During an overhaul the piston rings are replaced and the cylinder walls are bored and/or honed. If a rebore is done, then new pistons are also required. The main and connecting rod bearings are generally replaced with new ones and, if necessary, the crankshaft is also replaced. Generally the valves are serviced as well, since they are usually in less than perfect condition at this point. While the engine is being overhauled, other components such as the carburetors and the starter motor can be rebuilt also. The end result should be a like-new engine that will give as many trouble free miles as the original.
7   Before beginning the engine overhaul, read through all of the related procedures to familiarize yourself with the scope and requirements of the job. Overhauling an engine is not all that difficult, but it is time consuming. Plan on the motorcycle being tied up for a minimum of two (2) weeks. Check on the availability of parts and make sure that any necessary special tools, equipment and supplies are obtained in advance.
8   Most work can be done with typical shop hand tools, although a number of precision measuring tools are required for inspecting parts to determine if they must be replaced. Often a dealer service department or motorcycle repair shop will handle the inspection of parts and offer advice concerning reconditioning and replacement. As a general rule, time is the primary cost of an overhaul so it doesn't pay to install worn or substandard parts.
9   As a final note, to ensure maximum life and minimum trouble from a rebuilt engine, everything must be assembled with care in a spotlessly clean environment.

3 Repair operations requiring removal of the engine from the frame

1   Although many repair operations can be performed with the engine in place in the frame, others (especially those related to the lower end and the transmission) require that it be removed. Fortunately the removal procedure is not very difficult or time consuming.
2   Inspection and repair or replacement of the following components requires removal of the engine:
   Crankshaft/connecting rods
   Main and connecting rod bearings
   Rear cylinder/cylinder head components
   Transmission shafts/gears
   Shift drum/shift forks
3   Components/systems other than those mentioned are accessible for inspection and repair or replacement with the engine in the frame.

4   Engine — removal and installation

Note: *Engine removal and installation should be done with the aid of an assistant to avoid damage that could occur if the engine falls or is* *dropped. An hydraulic floor-type jack should be used to support and lower the engine if possible (they can be rented at low cost).*

Removal

1   Place the motorcycle on the centerstand, then refer to Chapter 1 and drain the engine oil and coolant.
2   Refer to Chapter 4 and remove the fuel tank and airbox. Disconnect the spark plug wires from the plugs.
3   Refer to Chapter 3 and remove the radiator.
4   Disconnect the cables from the battery (negative first, then positive). Detach the battery cable from the starter motor terminal, then work the cable free from the engine and bend it back along the right side of the frame.
5   Remove the exhaust pipe flange bolts from the front cylinder head (photo), then loosen the muffler clamp bolts and pull the pipes out of the front cylinder head and the muffler assembly. Store the pipes where they will not get dented or scratched.
6   Loosen the rear cylinder muffler clamp bolt, then support the muffler assembly and remove the passenger footpegs (one bolt each). Separate the muffler assembly from the rear cylinder exhaust pipe, then carefully withdraw it by passing it under the engine.
7   Refer to Chapter 3 and remove the coolant reservoir. Refer to Chapter 5 and remove the TCI unit and bracket.
8   Disconnect the alternator stator coil, pickup coil and sidestand

switch wiring connectors, then pass the wires between the frame and the battery case (you may have to remove the regulator/rectifier mounting screws to make room for the plastic wiring connectors to clear it).
9   Detach the negative battery cable from the left side engine case, then disconnect the wires from the coolant temperature sending unit and the thermostatic switch.
10   Refer to Chapter 4 and separate the choke and throttle cables from the carburetor. Disconnect the drain line from the fitting on the front carburetor. Separate the vent tubes from the fittings on top of the carburetors (photo).
11   Separate the clutch cable from the engine lever and remove it from the crankcase cover mount.
12   Remove the bolt and separate the rear brake pedal from the shaft.
13   Remove the bolt and separate the shift lever from the engine shaft. Separate the U-joint boot from the engine flange.
14   Remove the footpegs and the sidestand bracket. Leave the large stud that passes through the frame in place.
15   Remove the bolts and separate the bracket from the rear cylinder head and the frame.
16   Remove the right side frame tube (photo), then support the engine carefully on the floor jack (place a flat piece of wood between the jack and the engine crankcase to prevent damage) (photo).
17   Remove the large through-bolt (upper) and the left side frame bolts, then carefully pivot the engine forward and down on the large stud

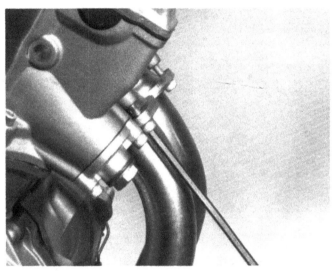

4.5   Removing the exhaust pipe flange bolts from the front cylinder head

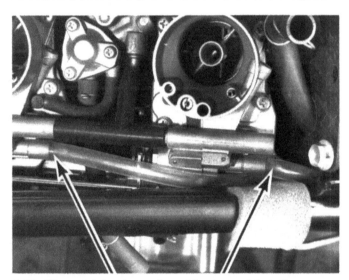

4.10   The vent tubes (arrows) must be removed from the carburetors and positioned out of the way

4.16a   The right side frame tube is held in place with five bolts and one nut (arrows)

4.16b   A floor jack should be used when removing the engine because it allows the engine to be lowered and moved ahead at the same time

(lower) by slowly lowering the jack.

18 As soon as the driveshaft is separated from the engine output shaft, support the engine and drive out the large stud. Carefully lower the engine (watch the carburetors) until it clears the frame, then move it forward and out the right side opening.

19 Separate the carburetors and the coolant hose assembly from the engine and plug the openings with clean paper towels or rags. Remove the exhaust pipes from the rear cylinder head.

Installation

20 Installation is basically the reverse of removal. Note the following important points:

a) Make sure the splines in the driveshaft line up properly with the splines on the output shaft.

b) Do not tighten any of the engine mounting bolts until they have all been installed.

c) When installing the sidestand, be careful not to pinch the side-stand switch wires between the sidestand bracket and the engine or frame.

d) Make sure the large lower stud is centered before tightening the footpeg nuts.

e) Use new gaskets at all exhaust pipe connections.

f) Route the clutch cable behind the left side of the radiator (adjust the clutch by referring to Chapter 1).

g) Make sure all wires behind the rear cylinder are clear of the exhaust pipe.

---

5 Engine disassembly and reassembly — general note

---

1 Before disassembling the engine, clean the exterior with a degreaser and rinse it with water. A clean engine will make the job easier and prevent the possibility of getting dirt into the internal areas of the engine.

2 In addition to the precision measuring tools mentioned earlier, you will need a torque wrench, a valve spring compressor, oil line brushes, a motorcycle piston ring removal and installation tool and a 46 millimeter (1-13/16 inch) deep socket or crowfoot wrench. Some new, clean engine oil of the correct grade and type, some engine assembly lube (or grease) with a high percentage of molybdenum disulfide and a tube of Yama-bond 4 sealant will also be required. Although it may not be considered a tool, some plastigage (type HPG-1) should also be obtained to use for checking bearing oil clearances (photos).

3 An engine support stand made from short lengths of 2 x 4's bolted together will facilitate the disassembly and reassembly procedures (photo).

4 When disassembling the engine, keep "mated" parts together (including gears, cylinders, pistons, etc. that have been in contact with each other during engine operation). These "mated" parts must be

5.2a A selection of brushes is required for cleaning holes and passages in the engine components

5.2b Yamabond 4 is used to seal the crankcase mating surfaces when they are rejoined

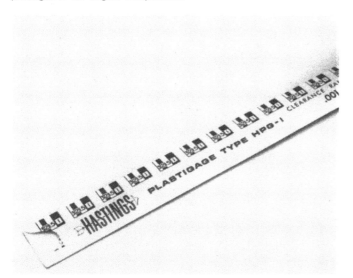

5.2c Type HPG-1 Plastigage is needed to check the connecting rod and camshaft bearing oil clearances

5.3 An engine stand can be made from short lengths of 2 x 4 lumber and lag bolts or nails

reused or replaced as an assembly.

5   Engine/transmission disassembly should be done in the following general order with reference to the appropriate Sections.

Remove the cylinder head(s)
Remove the right crankcase cover
Remove the cylinder(s) and piston(s)
Remove the clutch
Remove the oil pump
Remove the gearshift mechanism
Remove the alternator/starter clutch assembly
Remove the camshaft drive mechanisms
Remove the balancer drive gear
Separate the crankcase halves
Remove the balancer shaft
Remove the crankshaft and connecting rods
Remove the shift drum/forks
Remove the transmission shafts/gears

6   Reassembly is accomplished by reversing the general disassembly sequence.

6   Camshafts — removal, inspection and installation

Removal

1   Remove the spark plugs and the cylinder head covers (held in place with four bolts each). Do not lose the rubber plugs in the camshaft bearing caps (they may stick to the cylinder head covers as they are removed).

2   Remove the threaded plug and the cover plate from the left engine crankcase cover. Slip an appropriate size socket (with breaker bar attached) over the bolt in the left end of the crankshaft and turn the bolt counterclockwise until the T mark on the flywheel is aligned with the index mark in the cover and the camshaft lobes in the front cylinder are pointing up (no pressure on the valve lifters). At this point, the number one piston is at Top Dead Center (TDC) on the compression stroke. If the camshaft lobes are not pointing up, turn the crankshaft exactly 360° (counterclockwise) and align the marks as described previously.

3   Loosen the cap, then remove the cam chain tensioner from the cylinder head (it is held in place with two bolts) (photo).

4   Hold the camshaft with a 22 millimeter wrench and loosen the cam sprocket bolts (the bolts are extremely tight, so use a box end wrench or a 1/2 inch drive socket and breaker bar) (photo).

5   Stuff clean rags or paper towels into the cam chain tower in the cylinder head to keep parts from falling into the lower end of the engine, then carefully slip the sprockets and chain off of the camshafts and separate the sprockets from the chain. Let the chain roll up in the bottom of the chain tower. Label the sprockets so they can be reinstalled on the same camshaft they were removed from. **Note:** *Do not turn the*

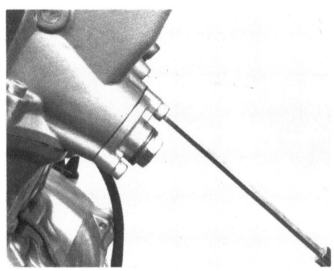

6.3   Removing the front cylinder cam chain tensioner bolts

6.4   Loosening the cam sprocket mounting bolts (note the 22 mm wrench position on the camshaft)

6.6a   Withdrawing the cam sprocket dowel pin

6.6b   Loosening the camshaft bearing cap bolts

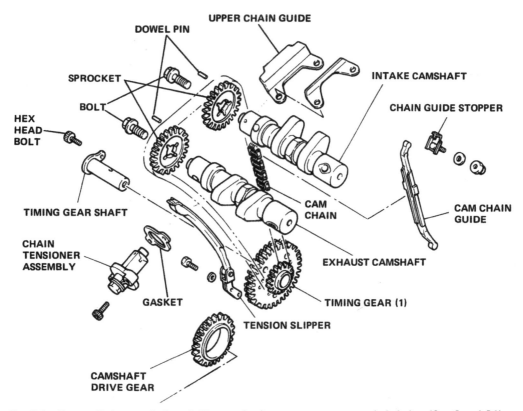

DOWEL PIN

UPPER CHAIN GUIDE

SPROCKET

INTAKE CAMSHAFT

BOLT

CHAIN GUIDE STOPPER

HEX HEAD BOLT

TIMING GEAR SHAFT

CAM CHAIN

CAM CHAIN GUIDE

CHAIN TENSIONER ASSEMBLY

EXHAUST CAMSHAFT

GASKET

TIMING GEAR (1)

TENSION SLIPPER

CAMSHAFT DRIVE GEAR

Fig. 2.1 *Front* cylinder camshaft and drive mechanism components — exploded view (Sec 6 and 21)

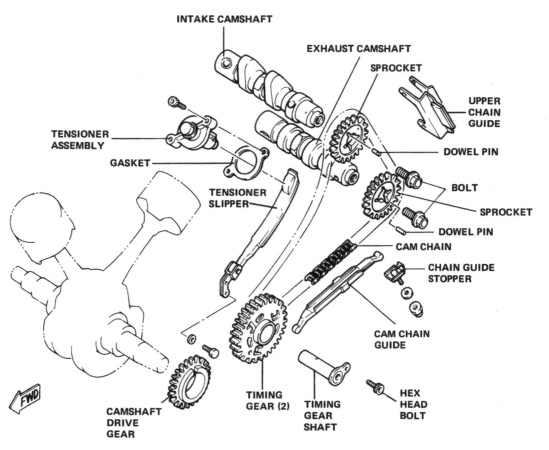

INTAKE CAMSHAFT

EXHAUST CAMSHAFT

SPROCKET

UPPER CHAIN GUIDE

TENSIONER ASSEMBLY

DOWEL PIN

GASKET

BOLT

TENSIONER SLIPPER

SPROCKET

DOWEL PIN

CAM CHAIN

CHAIN GUIDE STOPPER

CAM CHAIN GUIDE

FWD

CAMSHAFT DRIVE GEAR

TIMING GEAR (2)

TIMING GEAR SHAFT

HEX HEAD BOLT

Fig. 2.2 *Rear* cylinder camshaft and drive mechanism components — exploded view (Sec 6 and 21)

crankshaft with the chain and sprockets separated from the camshafts.

6   Carefully remove the dowel pins from the ends of the camshafts (photo), then gradually loosen the camshaft bearing cap bolts (1/8 turn each), following a criss-cross pattern (photo).

7   Remove the bolts and lift off the caps (label the caps to make sure they will be reinstalled in their original locations).

8   Lift out the camshafts and label them to ensure reinstallation in their original locations (they must not be mixed up).

Inspection

9   Inspect the cam bearing surfaces of the head and the bearing caps. Look for score marks, deep scratches and evidence of galling.

10   Check the camshaft lobes for heat discoloration (blue appearance), score marks, chipped areas and flat spots. Measure the height of each lobe (photo) and compare the results to the Specifications. Have the camshaft runout checked by a dealer service department or an automotive machine shop. If damage or excessive wear is evident, the camshaft must be replaced with a new one.

11   Next, check the camshaft bearing oil clearances. Clean the camshafts and the bearing surfaces in the cylinder head and bearing caps with a clean, lint-free cloth, then lay the cam in place in the cylinder head.

12   Cut two strips of Plastigage (type HPG-1) and lay one piece on each of the cam bearing journals, parallel with the camshaft centerline (photo).

13   Very carefully lay the cam bearing caps in place (the arrows on the caps must point toward the sprocket end of the camshaft) and install the bolts. Tighten the bolts to the specified torque in two steps, following a criss-cross pattern. *During this procedure do not rotate the camshaft.*

14   Loosen and remove the bolts (again, in two steps), then very carefully lift off the bearing caps.

15   To obtain the oil clearance, compare the crushed Plastigage (at its widest point) on each journal to the scale printed on the Plastigage container (photo). Compare the results to the Specifications. If the oil clearance is greater than specified, measure the cam bearing journal diameter with a micrometer (photo). If the journal diameter is less than the specified limit, replace the camshaft with a new one and recheck the clearance. If the clearance is still too great, replace the cylinder head and bearing caps with new parts.

16   Except in cases of oil starvation, the cam chain wears very little. If the chain has stretched excessively, which makes it difficult to maintain proper tension, replace it with a new one (Section 21).

17   Check the sprockets for wear, cracks and other damage. Replace questionable parts with new ones. **Note:** *If the sprockets are worn, the chain and drive sprocket are also worn. Replacement of these parts requires removal of the cylinder and head as well as the engine crankcase covers and possibly the alternator. Always replace the chain and all three sprockets for each cylinder when any one or all of the parts are worn or damaged.*

6.10   Measuring cam lobe height with a micrometer

6.12   Position the Plastigage strips (arrows) on the cam bearing journals, parallel with the centerline

6.15a   Comparing the width of the crushed Plastigage to the scale on the container to obtain the clearance

6.15b   Measuring the camshaft bearing journal diameter with a micrometer

18   Check the chain guides for wear and damage. If the guides are worn or damaged, the cam chain is worn out or improperly adjusted. Replacement of the slipper-type guides requires removal of the cylinder head, crankcase covers and the alternator (depending on which cylinder is involved).

Installation

19   Make sure the bearing surfaces in the cylinder head and the bearing caps are clean, then apply a light coat of engine assembly lube or molybdenum disulfide grease to each of them.

20   Make sure the camshaft bearing journals are clean, then lay the camshafts in the cylinder head (do not mix them up). Carefully set the bearing caps and upper chain guide in place (arrows pointing toward the sprocket end of the camshafts) and install the bolts. Tighten them gradually, in two steps, to the specified torque. Follow a criss-cross pattern to avoid warping the bearing caps.

Front cylinder

21   Turn the camshafts by hand and line up the punch marks with the index marks on the bearing caps. The cam lobes will be pointing up and toward each other (photo).

22   Install the dowel pins in the ends of the camshafts. **Caution:** *Do not drop them into the engine — extensive disassembly will be required to retrieve them.*

23   Make sure the timing marks are aligned as described in Paragraph 2, then mesh the intake camshaft sprocket with the chain and slip it over the end of the cam. The F mark must face out and the dowel pin must be engaged in the slot marked with an I (photo).

24   Mesh the exhaust camshaft sprocket with the chain and slip it into place on the end of the exhaust camshaft. The F mark must face out and the dowel pin must be engaged in the slot marked with an E (photo).

25   Insert your finger into the cam chain tensioner hole and apply

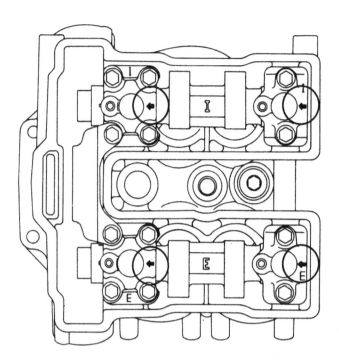

Fig. 2.3   Correct camshaft bearing cap installation (arrows pointing toward sprocket end of camshafts) (Sec 6)

CAM SPROCKET

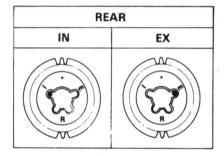

Fig. 2.4   Camshaft sprocket installation details (Sec 6)

6.21   Align the punch marks on the front cylinder cams with the index marks on the bearing caps (arrows); note the positions of the dowel pin holes

6.23   Correct location of dowel pin and F mark (arrows) when installing the *front* cylinder *intake* cam sprocket

pressure to the cam chain. Check the timing marks to make sure they are aligned (paragraph 2) and see if the punch marks on the camshaft are aligned with the index marks on the bearing caps. If necessary, change the position of the sprocket(s) on the chain to bring all of the marks into alignment. **Caution:** *If the marks are not aligned exactly as described, the valve timing will be incorrect.*

26   Install the sprocket mounting bolts and tighten them to the specified torque while holding the camshaft(s) with the 22 millimeter wrench.

27   Remove the cap and spring from the cam chain tensioner and depress the cam lock on the tensioner body while pushing the rod in.

28   Install the tensioner (use a new gasket) and tighten the mounting bolts to the specified torque, then install the spring and cap. When the cap is installed and tightened, the chain will automatically be tensioned correctly.

Rear cylinder

29   Turn the crankshaft exactly 290° in a counterclockwise direction (pull up on the chain as the crankshaft is turned) and align the 2T mark on the alternator rotor with the index mark in the crankcase cover.

30   The rear cylinder camshafts are installed in the same manner as the front cylinder camshafts, but note that the sprockets must be installed with the R (for rear) facing out (photos).

31   After the camshafts have been installed, coat the cam lobes with engine assembly lube or molybdenum disulfide grease (photo) and pour

clean engine oil of the recommended grade and type over the chains and sprockets. Turn the crankshaft through three or four revolutions (counterclockwise direction) and make sure that it turns smoothly and easily.

32   Make sure the rubber plugs are in place in the bearing caps, then install the cylinder head covers and tighten the bolts evenly and securely. **Caution:** *If the rubber plugs are left out, severe damage will occur to the camshafts, cylinder head and valve gear.*

33   Install the spark plugs.

34   After the engine has been run initially, refer to Chapter 1 and check/adjust the valve clearances.

---

7   Cylinder heads — removal and installation

Note: *Before attempting to remove the cylinder head(s), the cooling system, including the cylinder(s), must be drained completely. Refer to Chapter 1 for the procedure to follow.*

Removal

1   Refer to Section 6 and separate the chains and sprockets from the camshafts. **Note:** *It is not necessary to remove the camshafts from the cylinder heads when separating the heads from the cylinders, but if the engine is being completely disassembled, it is easier to loosen the*

6.24   Correct location of dowel pin and F mark (arrows) when installing the *front* cylinder *exhaust* cam sprocket

6.30a   Align the punch marks on the *rear* cylinder cams with the index marks on the bearing caps (arrows); note the positions of the dowel pin holes

6.30b   Correct location of dowel pin and R mark (arrows) when installing the *rear* cylinder *intake* cam sprocket

6.30c   Correct location of dowel pin and R mark (arrows) when installing the *rear* cylinder *exhaust* cam sprocket

cam bearing cap bolts while the heads are still attached to the cylinders.
2   Remove the rubber plugs from the cam bearing caps, then turn the camshafts by hand until the holes in the camshafts are aligned with the holes in the bearing caps.
3   Slip an extra long eight (8) millimeter hex (Allen) head socket wrench through the holes and engage it in the head nuts to loosen them (photo). Loosen the nuts gradually (about 1/8 of a turn at a time), in a criss-cross pattern, to avoid warping the head.
4   Remove the two small bolts from the side of the cylinder head (photo).
5   Carefully separate the head from the cylinder. You may have to tap lightly on the head with a soft-faced hammer to break the gasket seal, but do not pry between the cylinder and head to separate them (damage to the gasket sealing surfaces will result).
6   Note how they are installed, then remove the dowel pins and the large O-ring. Peel up the old head gasket and discard it (a new one must be used during installation).
7   Using a blunt gasket scraper or similar tool, remove all traces of old gasket material left on the head and cylinder gasket sealing surfaces. Clean them with a solvent such as lacquer thinner or acetone to ensure that the new gasket will adhere and seal properly.

Installation

8   Install the dowel pins and a new O-ring then lay the new head gasket in place (photo). *Never reuse the old gasket and do not use any type*

of gasket sealer.
9   Apply a small amount of engine assembly lube or molybdenum disulfide grease to the threads of the cylinder head studs.
10   Carefully set the cylinder head in place while directing the cam chain guides into their seats.
11   Tighten the head nuts gradually, following a criss-cross pattern, until the specified torque is reached. Don't forget to tighten the two small bolts as well.
12   Refer to Section 6 and attach the cam chain and sprockets to the camshafts.

8   Valves/valve seats/valve guides — servicing

1   Because of the complex nature of this job and the special tools and equipment required, servicing of the valves, the valve seats and the valve guides (commonly known as a valve job) is best left to a professional.
2   The home mechanic can, however, remove and disassemble the head, do the initial cleaning and inspection, then reassemble and deliver the head to a dealer service department or properly equipped motorcycle repair shop for the actual valve servicing. Refer to Section 9 for those procedures.
3   The dealer service department will remove the valves and springs, recondition or replace the valves and valve seats, replace the valve

6.31   Coat the camshaft lobes with engine assembly lube or molybdenum disulfide grease and make sure the rubber plugs are in place before installing the covers

7.3   Removing the head nuts with a long 8 mm hex (Allen) head socket and breaker bar

7.4   Removing the small cylinder head bolts

7.8   Correct installation of the dowel pins, O-ring and head gasket (rear cylinder shown)

guides, check and replace the valve springs, spring retainers and keepers (as necessary), replace the valve seals with new ones and reassemble the valve components.

4   After the valve job has been performed, the head will be in like-new condition. When the head is returned, be sure to clean it again very thoroughly before installation on the engine to remove any metal particles or abrasive grit that may still be present from the valve service operations. Use compressed air, if available, to blow out all the holes and passages.

5   **Caution:** *If valve service work is done, be sure to check the valve clearances (Chapter 1) before starting the engine.*

9   Cylinder head and valves — disassembly, inspection and reassembly

1   As was mentioned in the previous Section, valve servicing and valve guide replacement should be left to a dealer service department or motorcycle repair shop. However, disassembly, cleaning and inspection of the valves and related components can be done (if the necessary special tools are available) by the home mechanic. This way no expense is incurred if the inspection reveals that service work is not required at this time.

2   To properly disassemble the valve components without the risk of damaging them, a valve spring compressor is absolutely necessary. If the special tool is not available, have a dealer service department or motorcycle repair shop handle the entire process of disassembly, inspection, service or repair (if required) and reassembly of the valves.

Disassembly

3   Carefully remove the lifters from the cylinder head by inserting a small screwdriver into the slot and under the shim (photo). The shim should stay in place and the lifter should slide out easily. Label the lifters to make sure they are returned to their original bores during reassembly, then store them where they will not be scratched or otherwise damaged. **Note:** *Do not separate the shims from the lifters unless the lifters are to be replaced with new parts.*

4   Before the valves are removed, scrape away any traces of gasket material from the head gasket sealing surface. *Work slowly and do not nick or gouge the soft aluminum of the head.* Gasket removing solvents, which work very well, are available at most motorcycle shops and auto parts stores.

5   Carefully scrape all carbon deposits out of the combustion chamber area. A hand held wire brush or a piece of fine emery cloth can be used once the majority of deposits have been scraped away. Do not use a wire brush mounted in a drill motor, as the head material is soft and may be eroded away by the wire brush.

6   Before proceeding, arrange to label and store the valves along with their related components so they can be kept separate and reinstalled in the same valve guides they are removed from.

9.3   Sliding out a valve lifter with a small screwdriver

9.7a   Compressing the valve springs with a valve spring compressor

9.7b   Removing the keepers with a small pliers

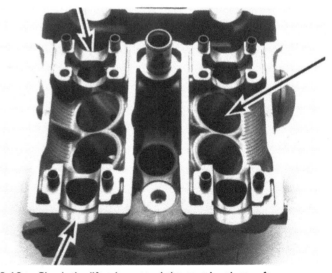

9.13a   Check the lifter bores and the cam bearing surfaces (arrows) in the cylinder head(s) for wear and damage

7   Compress the valve spring on the first valve with a spring compressor, then remove the keepers (photos) and the retainer from the valve assembly. Do not compress the springs any more than is absolutely necessary. Carefully release the valve spring compressor and remove the springs and the valve from the head. If the valve binds in the guide (won't pull through), push it back into the head and deburr the area around the keeper groove with a very fine file or whetstone.

8   Repeat the procedure for the remaining valves. Remember to keep the parts for each valve together so they can be reinstalled in the same location.

9   Once the valves have been removed and labeled, pull off the valve stem seals with pliers and discard them *(the old seals should never be reused)*, then remove the spring seats.

10   Next, clean the cylinder head with solvent and dry it thoroughly. Compressed air will speed the drying process and ensure that all holes and recessed areas are clean.

11   Clean all of the valve springs, keepers, retainers and spring seats with solvent and dry them thoroughly. *Do the parts from one valve at a time so that no mixing of parts between valves occurs.*

12   Scrape off any deposits that may have formed on the valve, then use a motorized wire brush to remove deposits from the valve heads and stems. Again, make sure the valves do not get mixed up.

Inspection

13   Inspect the head very carefully for cracks and other damage. If cracks are found, a new head is in order. Check the lifter bores and the cam bearing surfaces for wear and evidence of seizure (photo). Check the lifters for wear as well (photo).

14   Using a straightedge and feeler gauge, check the head gasket

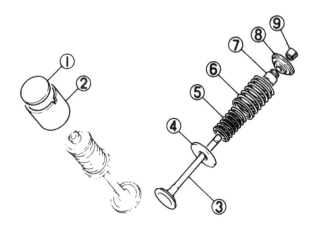

Fig. 2.5   Valve components — exploded view (Sec 9)

1   Shim
2   Lifter
3   Valve
4   Spring seat
5   Inner spring
6   Outer spring
7   Valve stem-to-guide seal
8   Retainer
9   Keepers

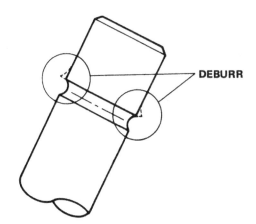

DEBURR

Fig. 2.6   If the valve binds in the guide, deburr the area above the keeper groove (Sec 9)

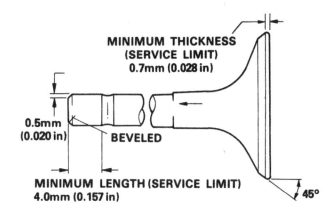

MINIMUM THICKNESS (SERVICE LIMIT) 0.7mm (0.028 in)

0.5mm (0.020 in)   BEVELED

MINIMUM LENGTH (SERVICE LIMIT) 4.0mm (0.157 in)   45°

Fig. 2.7   Valve inspection details (Sec 9)

9.13b   Check the lifters for scuffing and score marks

9.14   Checking a cylinder head for warpage with a straightedge (dial caliper body) and a feeler gauge

mating surface for warpage. Lay the straightedge lengthwise, across the head and diagonally (corner-to-corner), intersecting the head bolt holes, and try to slip a 0.0012 in (0.03 mm) feeler gauge under it at each location (photo). If the feeler gauge can be inserted between the head and the straightedge, the head is warped and must be replaced with a new one.

15   Examine the valve seats in each of the combustion chambers. If they are pitted, cracked or burned, the head will require valve service that is beyond the scope of the home mechanic. Measure the valve seat width (photo) and compare it to the Specifications. If it is not within the specified range, or if it varies around its circumference, valve service work is required.

16   Clean the valve guides to remove any carbon buildup, then measure the inside diameters of the guides (at both ends and the center of the guide) with a small hole gauge and a 0-to-1 inch micrometer (photos). Record the measurements for future reference. These measurements, along with the valve stem diameter measurements, will enable you to compute the valve stem-to-guide clearance. This clearance, when compared to the Specifications, will be one factor that will determine the extent of the valve service work required. The guides are measured at the ends and at the center to determine if they are worn in a bell mouth pattern (more wear at the ends). If they are, guide replacement is an absolute must.

17   Carefully inspect each valve face for cracks, pits and burned spots.

Check the valve stem and the keeper groove area for cracks (photo). Rotate the valve and check for any obvious indication that it is bent. Check the end of the stem for pitting and excessive wear and make sure the bevel is the specified width. The presence of any of the above conditions indicates the need for valve servicing.

18   Measure the valve stem diameter (photo). By subtracting the stem diameter from the valve guide diameter, the valve stem-to-guide clearance is obtained. If the stem-to-guide clearance is greater than specified, the guides and valves will have to be replaced with new ones.

19   Check the end of each valve spring for wear and pitting. Measure the free length (photo) and compare it to the Specifications. Any springs that are shorter than specified have sagged and should not be reused. Stand the spring on a flat surface and check it for squareness (photo).

20   Check the spring retainers and keepers for obvious wear and cracks. Any questionable parts should not be reused, as extensive damage will occur in the event of failure during engine operation.

21   If the inspection indicates that no service work is required, the valve components can be reinstalled in the head.

Reassembly

22   Before installing the valves in the head, they should be lapped to ensure a positive seal between the valves and seats. This procedure requires fine valve lapping compound (available at auto parts stores) and a valve lapping tool. If a lapping tool is not available, a piece of

9.15   Measuring the valve seat width with a ruler

9.16a   Using a small hole gauge to determine the diameter of the valve guide

9.16b   Measuring the small hole gauge with a micrometer to obtain the actual size of the valve guide in inches

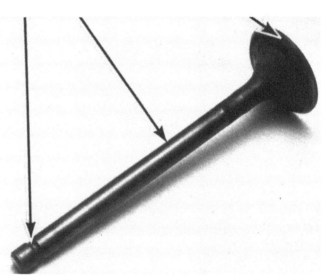

9.17   Check the valve face, stem and keeper groove (arrows) for signs of wear and damage

rubber or plastic hose can be slipped over the valve stem (after the valve has been installed in the guide) and used to turn the valve.

23   Apply a small amount of fine lapping compound to the valve face (photo), then slip the valve into the guide. **Note:** *Make sure the valve is installed in the correct guide and be careful not to get any lapping compound on the valve stem.*

24   Attach the lapping tool (or hose) to the valve and rotate the tool between the palms of your hands. Use a back-and-forth motion rather than a circular motion (photo). Lift the valve off the seat at regular intervals to distribute the lapping compound properly (photo). Continue the lapping procedure until the valve face and seat contact area is of uniform width and unbroken around the entire circumference of the valve face and seat (photos).

25   Carefully remove the valve from the guide and wipe off all traces of lapping compound. Use solvent to clean the valve and wipe the seat area thoroughly with a solvent soaked cloth. Repeat the procedure for the remaining valves.

26   Lay the spring seats in place in the cylinder head, then install new valve stem seals on each of the guides. Use an appropriate size deep socket to push the seals into place until they are properly seated. Do not twist or cock them or they will not seal properly against the valve stems and do not remove them again or they will be damaged.

27   Next, install the valves (taking care not to damage the new seals), the springs, the retainers and the keepers. Coat the valve stems with

assembly lube or grease (preferably molybdenum disulfide) before slipping them into the guides and install the springs with the tightly wound coils next to the cylinder head (photo). When compressing the springs with the valve spring compressor, depress them only as far as is absolutely necessary to slip the keepers into place. Apply a small amount of grease to the keepers (photo) to help hold them in place as the pressure is released from the springs. Make certain that the keepers are securely locked in their retaining grooves.

28   Support the cylinder head on blocks so the valves cannot contact the workbench top and very gently tap each of the valve stems with a soft-faced hammer. This will help seat the keepers in their grooves.

29   Once all of the valves have been installed in the head, check for proper valve sealing by pouring a small amount of solvent into each of the valve ports. If the solvent leaks past the valve(s) into the combustion chamber area, disassemble the valve(s) and repeat the lapping procedure, then reinstall the valve(s) and repeat the check. Repeat the procedure until a satisfactory seal is obtained.

10   Cylinders — removal, inspection and installation

1   Refer to the appropriate Section and remove the cylinder head or heads. If both cylinders are being removed, mark them to ensure that they are reinstalled in the same location.

9.18   Measuring the valve stem diameter with a micrometer

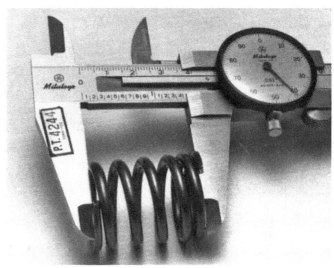

9.19a   Measuring the valve spring free length with a dial caliper

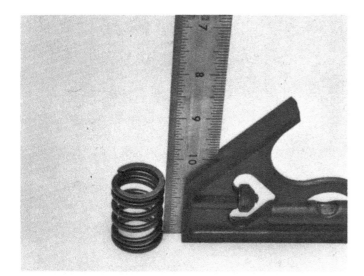

9.19b   Checking a valve spring for squareness

9.23   Apply the lapping compound very sparingly, in small dabs, to the valve *face* only

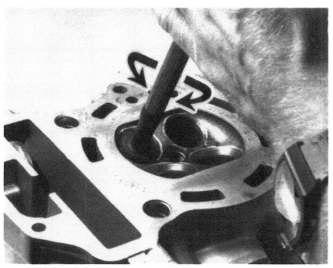

9.24a Rotate the valve lapping tool back-and-forth between the palms of your hands

9.24b Lift the tool and valve periodically to redistribute the lapping compound

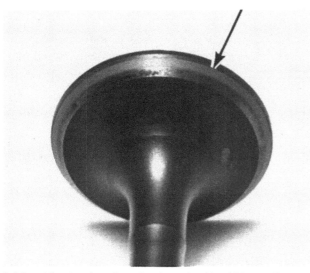

9.24c After lapping, the valve face should exhibit a uniform, unbroken contact pattern (arrow). . .

9.24d . . . and the seat should be the specified width (arrow), with a smooth, unbroken appearance

9.27a Install the springs with the closely spaced coils (arrow) next to the cylinder head

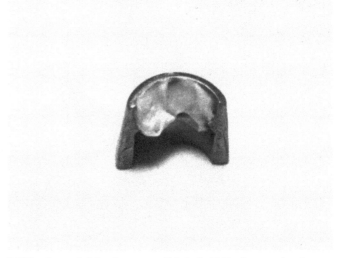

9.27b A small dab of grease will help hold the keepers in place on the valve while the spring is released

Removal

2   Remove the cam chain guide from the cylinder, then refer to Chapter 3 and remove the water pump cover and coolant tubes.

3   The cylinders will probably be stuck tightly to the crankcase, so tap around their entire circumference with a soft-faced hammer to break the gasket seal.

4   Rotate the crankshaft until the pistons are as far up in the bores as possible, then separate the cylinders from the crankcase.

5   Before the cylinders are lifted off the pistons, stuff a clean shop towel into each crankcase opening to keep foreign objects out of the engine and to cushion the pistons/connecting rods as the cylinders are removed.

6   Remove the O-ring and the dowel pins, then peel up the cylinder base gasket.

7   Remove the O-rings from the base of the cylinder sleeves. Using a blunt gasket scraper or similar tool, remove all traces of old gasket material left on the cylinders and crankcase, then clean the cylinders with solvent and dry them thoroughly. Do not separate the coolant tube elbows from the cylinders unless coolant has been leaking from the cylinder-to-elbow joint. The front cylinder elbow is marked with a 1; the rear cylinder elbow is marked with a 2.

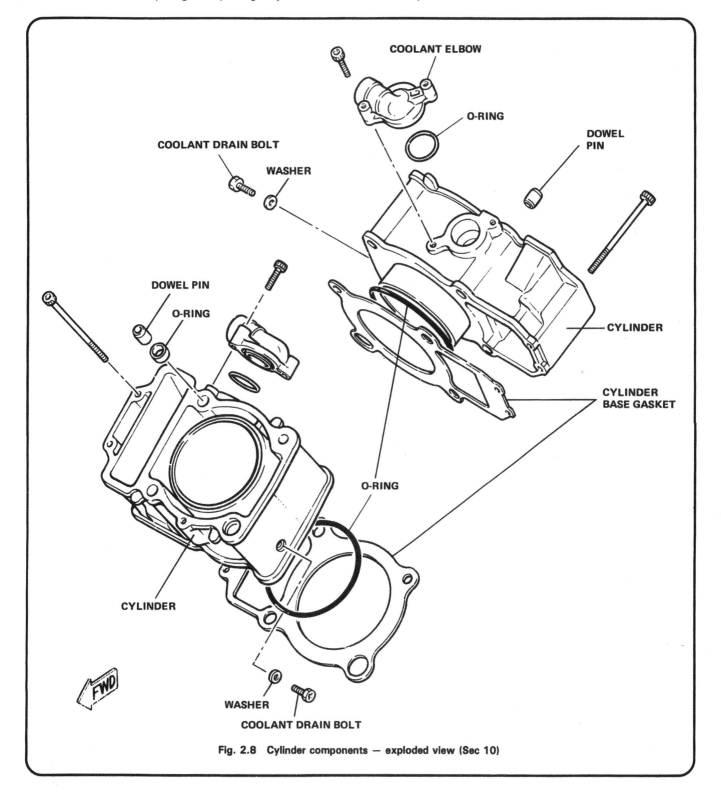

Fig. 2.8 Cylinder components — exploded view (Sec 10)

Inspection

8   Check the cylinder walls carefully for scratches and score marks.

9   Using the appropriate precision measuring tools, check each cylinder's diameter near the top, center and bottom of the cylinder bore, parallel to the crankshaft axis (photo). Next, measure each cylinder's diameter at the same three locations across the crankshaft axis. Compare the results to the Specifications. If the cylinder walls are tapered, out-of-round, worn beyond the specified limits, or badly scuffed or scored, have them rebored and honed by a dealer service department or a motorcycle repair shop. If a rebore is done, oversize pistons and rings will be required as well.

10   As an alternative, if the precision measuring tools are not available, a dealer service department or motorcycle repair shop will make the measurements and offer advice concerning servicing of the cylinders.

11   If they are in reasonably good condition and not worn to the outside of the limits, and if the piston-to-cylinder clearances can be maintained properly (Section 11), then the cylinders do not have to be rebored; honing is all that is necessary.

12   To perform the honing operation you will need the proper size flexible hone with fine stones, plenty of light oil or honing oil, some shop towels and an electric drill motor. Hold the cylinder in a vise (cushioned with soft jaws or wood blocks) when performing the honing operation. Mount the hone in the drill motor, compress the stones and slip the hone into the cylinder. Lubricate the cylinder thoroughly, turn on the drill and move the hone up and down in the cylinder at a pace which will produce a fine crosshatch pattern on the cylinder wall with the crosshatch lines intersecting at approximately a 60° angle. Be sure to use plenty of lubricant and do not take off any more material than is absolutely necessary to produce the desired effect. Do not withdraw the hone from the cylinder while it is running. Instead, shut off the drill and continue moving the hone up and down in the cylinder until it comes to a complete stop, then compress the stones and withdraw the hone. Wipe the oil out of the cylinder and repeat the procedure on the remaining cylinder. Remember, do not remove too much material from the cylinder wall. If you do not have the tools, or do not desire to perform the honing operation, a dealer service department or motorcycle repair shop will generally do it for a reasonable fee.

13   Next, the cylinders must be thoroughly washed with *warm soapy water* to remove all traces of the abrasive grit produced during the honing operation. Be sure to run a brush through the bolt holes and flush them with running water. After rinsing, dry the cylinders thoroughly and apply a coat of light, rust-preventative oil to all machined surfaces.

Installation

14   Before installation, clean the gasket sealing surfaces of the crankcase and the cylinders with a solvent such as lacquer thinner or acetone.

15   Make sure the oil control orifice is unobstructed and tight (photo), then install the three dowel pins and the O-ring in the crankcase recesses.

10.9   Checking the cylinder diameter with a telescoping gauge (the gauge is then measured with a micrometer)

10.15   Make sure the oil orifice (arrow) is clear and tight before installing the cylinder

10.16   Correct installed position of the dowel pins, O-ring and cylinder base gasket (rear cylinder shown)

10.17   Make sure the O-ring is in place on the cylinder (arrow) before installing the cylinder on the crankcase

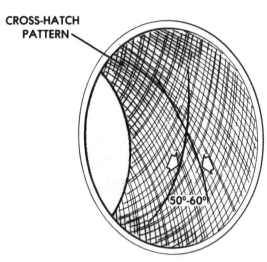

CROSS-HATCH
PATTERN

50°-60°

Fig. 2.9 The cylinder hone should leave a crosshatch pattern with the lines intersecting at approximately a 60 degree angle (Sec 10)

11.3 Removing the piston pin circlip (note the rag used to keep foreign objects out of the crankcase)

11.4 Use large rubber bands to keep the connecting rods from flopping around after the pistons are removed

16 Lay a new cylinder base gasket in place (photo) (no gasket sealer is necessary), then roll the camshaft drive chain up and position it in the cavity in front of the chain guide.
17 Install new O-rings on the cylinder sleeves (photo), then make sure the coolant drain bolts are tight. Apply a thin coat of clean engine oil to the cylinder bore.
18 Position the cylinder over the piston, hook the cam chain with a length of bent wire and pull it up through the chain tunnel. Direct the chain tensioner slipper into the tunnel as well.
19 Carefully lower the first cylinder down over the studs while directing the piston into the cylinder bore. The long tapered portion at the bottom of the bore will serve to compress the rings as the piston enters the bore (rock the piston slightly to help ease the rings into position). *If you encounter resistance, do not force the cylinder down, as the piston ring lands are easily broken.* Repeat the procedure for the remaining cylinder.
20 Refer to the appropriate Section and install the cylinder heads.

11 Pistons — removal, inspection and installation

1 The pistons are attached to the connecting rods with piston pins that are a slip fit in the pistons and rods.
2 Before removing the pistons from the rods, stuff a clean shop towel into each crankcase hole, around the connecting rods. This will prevent the circlips from falling into the crankcase if they are inadvertently dropped.

Removal
3 Mark the pistons with an F (front) and an R (rear) to ensure proper installation. Support the first piston, grasp the circlip with needlenose pliers and remove it from the groove (photo).
4 Push the piston pin out from the opposite end to free the piston from the rod. You may have to deburr the area around the groove to enable the pin to slide out (use a triangular file for this procedure). Repeat the procedure for the remaining piston. Use large rubber bands to support the connecting rods (photo).

Inspection
5 Before the inspection process can be carried out, the pistons must be cleaned and the old piston rings removed.
6 Using a piston ring installation tool, carefully remove the rings from the pistons (photo). Do not nick or gouge the pistons in the process.
7 Scrape all traces of carbon from the tops of the pistons. A handheld wire brush or a piece of fine emery cloth can be used once the majority of the deposits have been scraped away. Do not, under any circumstances, use a wire brush mounted in a drill motor to remove deposits from the pistons; the piston material is soft and will be eroded away by the wire brush.

11.6 Removing a piston ring with a piston ring removal and installation tool

8   Use a piston ring groove cleaning tool to remove any carbon deposits from the ring grooves. If a tool is not available, a piece broken off the old ring will do the job. Be very careful to remove only the carbon deposits. Do not remove any metal and do not nick or gouge the sides of the ring grooves.

9   Once the deposits have been removed, clean the pistons with solvent and dry them thoroughly. Make sure the oil return holes below the oil ring grooves are clear.

10   If the pistons are not damaged or worn excessively and if the cylinders are not rebored, new pistons will not be necessary. Normal piston wear appears as even, vertical wear on the thrust surfaces of the piston and slight looseness of the top ring in its groove. New piston rings, on the other hand, should always be used when an engine is rebuilt.

11   Carefully inspect each piston for cracks around the skirt, at the pin bosses and at the ring lands (photo).

12   Look for scoring and scuffing on the thrust faces of the skirt, holes in the piston crown and burned areas at the edge of the crown. If the skirt is scored or scuffed, the engine may have been suffering from overheating and/or abnormal combustion, which caused excessively high operating temperatures. The oil pump and cooling system should be checked thoroughly. A hole in the piston crown, an extreme to be sure, is an indication that abnormal combustion (pre-ignition) was occurring. Burned areas at the edge of the piston crown are usually evidnece of spark knock (detonation). If any of the above problems exist,

the causes must be corrected or the damage will occur again.

13   Measure the piston ring-to-groove clearance by laying a new piston ring in the ring groove and slipping a feeler gauge in beside it (photo). Check the clearance at three or four locations around the groove. Be sure to use the correct ring for each groove; they are different. If the clearance is greater then specified, new pistons will have to be used when the engine is reassembled.

14   Check the piston-to-bore clearance by measuring the bore (see Section 10) and the piston diameter. Make sure that the pistons and cylinders are correctly matched. Measure the piston across the skirt on the thrust faces at a 90° angle to the piston pin, about 1/2 inch (11 mm) up from the bottom of the skirt (photo). Subtract the piston diameter from the bore diameter to obtain the clearance. If it is greater than specified, the cylinders will have to be rebored and new oversized pistons and rings installed. If the appropriate precision measuring tools are not available, the piston-to-cylinder clearances can be obtained, though not quite as accurately, using feeler gauge stock. Feeler gauge stock comes in 12 inch lengths and various thicknesses and is generally available at auto parts stores. To check the clearance, select a 0.002 in (0.05 mm) feeler gauge and slip it into the cylinder along with the appropriate piston. The cylinder should be upside down and the piston must be positioned exactly as it normally would be. Place the feeler gauge between the piston and cylinder on one of the thrust faces (90° to the piston pin bore). The piston should slip through the cylinder (with the feeler gauge in place) with moderate pressure. If it falls through,

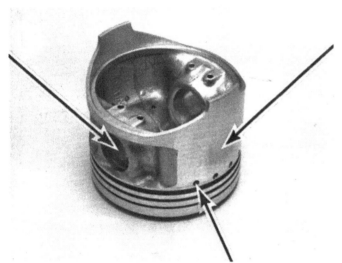

11.11   Check the piston pin bore and the piston skirt for wear and make sure the oil holes are clear (arrows)

11.13   Measuring piston ring-to-groove clearance with a feeler gauge

11.14   Measuring piston diameter with a micrometer

11.17   Make sure the EX mark is facing to the front on the front cylinder and to the rear on the rear cylinder

or slides through easily, the clearance is excessive and a new piston will be required. If the piston binds at the lower end of the cylinder and is loose toward the top, the cylinder is tapered, and if tight spots are encountered as the piston/feeler gauge is rotated in the cylinder, the cylinder is out-of-round. Repeat the procedure for the remaining piston and cylinder. Be sure to have the cylinders and pistons checked by a dealer service department or a motorcycle repair shop to confirm your findings before purchasing new parts.

15   Apply clean engine oil to the pin, insert it into the piston and check for free play by rocking the pin back-and-forth. If the pin is loose, new pistons and possibly new pins must be installed.

16   Refer to Section 12 and install the rings on the pistons.

Installation

17   Install the pistons in their original locations with the EX (exhaust) mark on the front piston facing forward and the mark on the rear piston

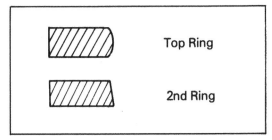

Fig. 2.10   Piston ring cross sections (note that the 2nd ring is actually thicker than the top ring) (Sec 12)

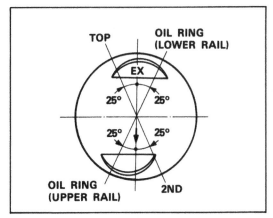

Fig. 2.11   Piston ring end gap positions before cylinder installation (Sec 12)

facing the rear of the engine (photo). Lubricate the pins and the rod bores with clean engine oil. Push the pins into position and install new circlips *(do not reuse the old clips)*. Make sure the clips are properly seated in the grooves, then turn them so the gaps are up.

12   Piston rings — installation

1   Before installing the new piston rings, the ring end gaps must be checked.

2   Lay out the pistons and the new ring sets so the rings will be matched with the same piston and cylinder during the end gap measurement procedure and engine assembly.

3   Insert the top (No. 1) ring into the bottom of the first cylinder and square it up with the cylinder walls by pushing it in with the top of the piston. The ring should be about 1 inch above the bottom edge of the cylinder. To measure the end gap, slip a feeler gauge between the ends of the ring (photo) and compare the measurement to the Specifications.

4   If the gap is larger or smaller than specified, double check to make sure that you have the correct rings before proceeding.

5   If the gap is too small, it must be enlarged or the ring ends may come in contact with each other during engine operation, which can cause serious damage. The end gap can be increased by filing the ring ends very carefully with a fine file. Mount the ring in a vise equipped with soft jaws, holding it as close to the gap as possible. When performing this operation, file only from the outside in.

6   Excess end gap is not critical unless it is greater than 0.040 in (1 mm). Again, double check to make sure you have the correct rings for your engine.

7   Repeat the procedure for each ring that will be installed in the first cylinder and for each ring in the remaining cylinder. Remember to keep the rings, pistons and cylinders matched up.

8   Once the ring end gaps have been checked/corrected the rings can be installed on the pistons.

9   The oil control ring (lowest on the piston) is installed first. It is composed of three separate components. Slip the spacer expander into the groove, then install the upper side rail. *Do not use a piston ring installation tool on the oil ring side rails as they may be damaged.* Instead, place one end of the side rail into the groove between the spacer expander and the ring land. Hold it firmly in place and slide a finger around the piston while pushing the rail into the groove. Next, install the lower side rail in the same manner.

10   After the three oil ring components have been installed, check to make sure that both the upper and lower side rails can be turned smoothly in the ring groove.

11   the No. 2 (middle) ring is installed next. It can be readily distinguished from the top ring by its cross section shape and the fact that it is thicker than the top ring. Do not mix the top and middle rings, as they have different cross sections.

12.3   Checking piston ring end gap with a feeler gauge

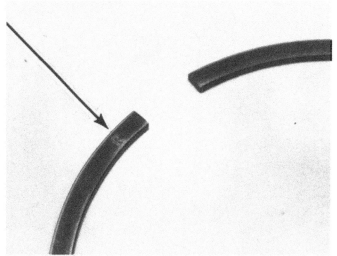

12.12   Make sure the marks on the rings (arrow) face up when the rings are attached to the pistons

12 To avoid breaking the ring, use a piston ring installation tool and make sure that the identification mark is facing up (photo). Fit the ring into the middle groove on the piston. Do not expand the ring any more than is necessary to slide it into place.

13 Finally, install the No. 1 (top) ring in the same manner. Make sure the identifying mark is facing up.

14 Repeat the procedure for the remaining piston and rings. Be very careful not to confuse the No. 1 and No. 2 rings.

15 Once the rings have been properly installed, turn them as required to position the end gaps as shown in the accompanying illustration.

13 Clutch — removal

1 It should be noted that the clutch assembly can be removed with the engine in the frame. To do so, first drain the engine oil and remove the thermostat housing (Chapter 3).

2 Remove the right crankcase cover bolts. It may be helpful to make a simple holder from a piece of cardboard to ensure that the bolts are returned to their original locations when the cover is installed.

3 Tap the crankcase cover gently with a soft-faced hammer to break

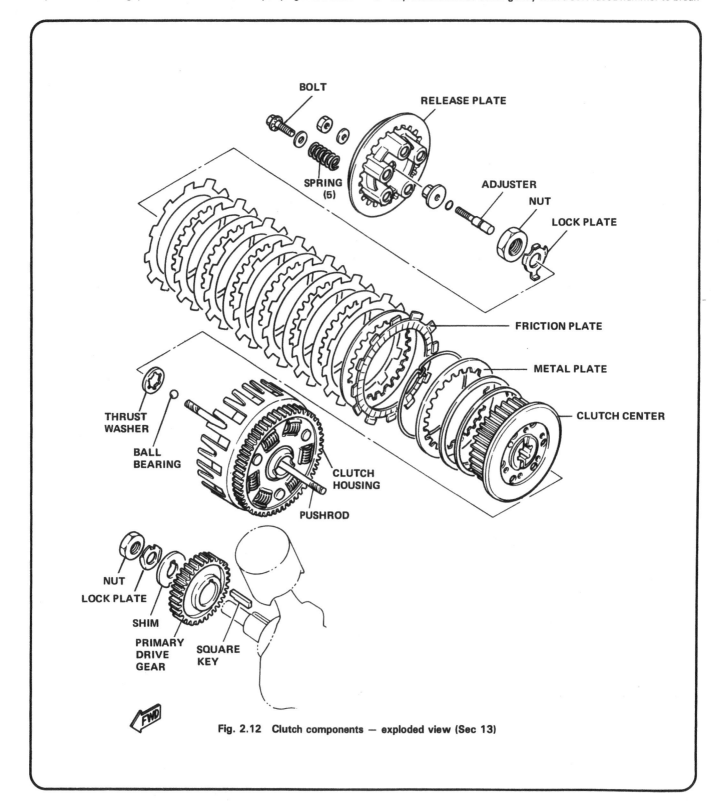

Fig. 2.12 Clutch components — exploded view (Sec 13)

the gasket seal, then pull it away from the engine. Do not pry between the gasket sealing surfaces, as damage and eventually oil leaks will occur.

4 Remove the clutch release plate by unscrewing the five mounting bolts (photo). Loosen them gradually, one turn at a time each, following a criss-cross pattern until the pressure from the springs has been released.

5 Remove the ball bearing and the clutch pushrod from the hole in the transmission mainshaft, pull out the clutch plates, bend back the lockplate tab, then loosen the clutch nut (photos). In order to do this, the transmission must be shifted into gear and the output shaft U-joint must be held with a large screwdriver or other suitable tool to keep it from turning as the nut is loosened. If the engine is in the frame, shift the transmission into gear and apply the rear brake as the nut is loosened.

6 Remove the nut and the lock plate, then pull the clutch center and housing off the shaft.

7 Refer to Section 14 for clutch component inspection procedures. If the engine is being totally disassembled, proceed to the oil pump removal procedure.

14 Clutch — inspection

1 The clutch center contains a built-in damper beneath the first steel

13.4 Loosening the clutch release plate bolts

13.5a Removing the ball bearing from the mainshaft bore with a magnet

13.5b Removing the clutch pushrod from the mainshaft bore

13.5c Bending back the lock plate tab with a cold chisel

14.1 Check the clutch center splines (arrows) for wear and evidence of distortion

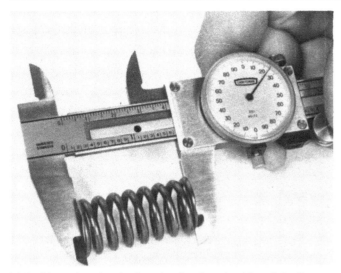

14.2　Measuring the clutch spring free length with a dial caliper

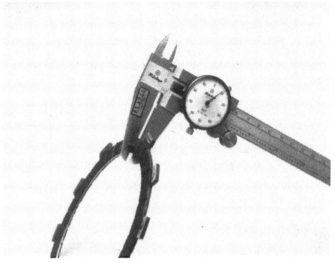

14.3　Measuring the thickness of a clutch friction plate with a dial caliper

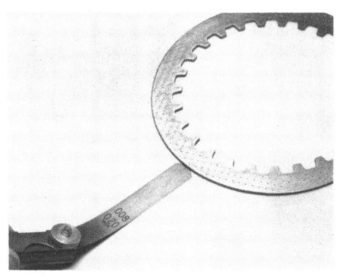

14.4　Checking a clutch metal plate for warpage

14.6　Check the slots in the clutch housing for indentations (minor damage can be removed with a file) and check the bearing for wear (arrows)

14.7a　Check the transmission mainshaft for wear and damage

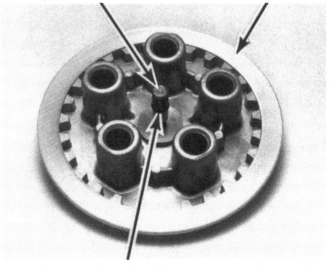

14.7b　Check the release plate friction surface, the adjuster and the O-ring (arrows) for wear and damage

plate. Do not remove the set ring and clutch plate unless abnormal clutch chatter has been evident. Examine the splines on both the inside and the outside of the clutch center (photo). If any wear is evident, replace the clutch center with a new one.

2 Measure the free length of the clutch springs (photo) and compare the results to the Specifications. If the springs have sagged, or if cracks are noted, replace them with new ones as a set.

3 If the lining material of the friction plates smells burnt or if it is glazed, new parts are required. If the metal clutch plates are scored or discolored, they must be replaced with new ones. Measure the thickness of each friction plate (photo) and compare the results to the Specifications. Replace with new parts any friction plates that are near the wear limit.

4 Lay the metal plates, one at a time, on a perfectly flat surface (such as a piece of plate glass) and check for warpage by trying to slip a 0.008 in feeler gauge between the flat surface and the plate (photo). Do this at several places around the plate's circumference. If the feeler gauge can be slipped under the plate, it is warped and should be replaced with a new one.

5 Check the tabs on the friction plates for excessive wear and mushroomed edges. They can be cleaned up with a file if the deformation is not severe.

6 Check the edges of the slots in the clutch housing for indentations made by the friction plate tabs (photo). If the indentations are deep they can prevent clutch release, so the housing should be replaced with a new one. If the indentations can be removed easily with a file, the life of the housing can be prolonged to an extent. Also, check the primary gear teeth for cracks, chips and excessive wear. If the gear is worn or damaged, the clutch housing must be replaced with a new one. Check the bearing for score marks, scratches and excessive wear.

7 Check the bearing journal on the transmission mainshaft for score marks, heat discoloration and evidence of excessive wear (photo). Check the clutch release plate, adjuster and O-ring for wear and damage and make sure the long pushrod is not bent (roll it on a perfectly flat surface or use V-blocks and a dial indicator).

8 Refer to Section 15 for clutch installation procedures.

15 Clutch — installation

1 Lubricate the thrust washer and the bearing journal on the transmission mainshaft with molybdenum disulfide grease, then slide the clutch housing into place, followed by the thrust washer (photo).

2 Install the clutch center on the mainshaft, then slip the lock plate into place and thread the large clutch nut onto the mainshaft. Note that the tangs on the lock plate must fit into the slots in the clutch center hub.

3 Tighten the nut to the specified torque, then bend the unused locking tab up against the flat of the nut. If both locking tabs have been used, install a new locking plate.

4 Install the clutch plates. Begin with a friction plate and alternate friction and metal plates until all of them are in place in the clutch assembly.

5 Lubricate the pushrod and install it in the mainshaft hole, followed by the ball bearing.

6 Lubricate the O-ring on the release plate adjuster, then install the plate by lining up the arrow on the plate with the indentation in the clutch center (photo).

7 Install the springs and bolts and tighten the bolts gradually, following a criss-cross pattern (to compress the springs evenly) to the specified torque.

8 Unless the engine is being disassembled, install the right crankcase cover and thermostat housing (Chapter 3), then adjust the clutch by referring to Chapter 1.

16 Oil pump — removal and disassembly

1 The oil pump is located behind the right crankcase cover, directly in front of the clutch. It should be noted that it is accessible for repair or replacement with the engine in the frame. In that case, preliminary disassembly includes draining the engine oil and removing the thermostat housing (Chapter 3).

2 Remove the right crankcase cover bolts and separate the cover from

15.1 Make sure the thrust washer is in place (arrow) before installing the clutch center

15.6 Align the marks (arrows) on the clutch center and the release plate during installation of the plate

16.4 Removing the snap-ring from the oil pump shaft

the engine. You may have to tap it gently with a soft-faced hammer to break the gasket seal.

3   If the engine is being completely disassembled, the clutch can be removed before the oil pump.

Removal

4   Remove the snap-ring from the oil pump shaft (photo), then slip the nylon gear off the shaft.

5   Remove the screws that attach the pump to the crankcase. They are extremely tight, so a breaker bar and hex (Allen) head socket should be used (photo). Note the O-ring in the crankcase hole.

Disassembly

6   Remove the single screw from the backside and separate the pump cover from the body.

7   Push the shaft and the inner rotor out of the pump body. The outer rotor will simply fall out of place; *do not drop it*. Do not separate the pickup tube from the pump unless the screen is clogged or damaged.

8   Refer to Section 17 for oil pump inspection procedures.

---

17   Oil pump — inspection

1   Clean the parts with solvent and dry them thoroughly. If available, use compressed air to blow out all of the cavities.

2   Check the entire pump body and cover for cracks and evidence of wear. Look closely for a ridge where the rotors contact the body and cover (photo).

3   Reassemble the rotors and the shaft in the pump body and check the inner rotor-to-outer rotor clearance and the outer rotor-to-pump body clearance (photos).

4   If the oil pump clearances are excessive, or if excessive wear is evident, replace the oil pump as a complete unit.

5   Refer to Section 18 for oil pump reassembly and installation procedures.

---

18   Oil pump — reassembly and installation

1   As the parts are assembled, lubricate them liberally with clean engine oil, a molybdenum disulfide based oil additive, or grease (preferably one containing molybdenum disulfide).

Reassembly

2   Install the outer rotor in the pump body. Slip the drive pin through the shaft, then slide the inner rotor onto the shaft and engage the slots in the rotor with the drive pin ends.

3   Insert the shaft through the pump body and mesh the rotors, then

16.5   Removing the oil pump mounting bolts

17.2   Check the oil pump for wear in the areas indicated by the arrows

17.3a   Checking the oil pump inner rotor-to-outer rotor clearance with a feeler gauge

17.3b   Checking the oil pump outer rotor-to-body clearance

install the cover (with the dowel pin in place) and tighten the screw securely.
4   Make sure the pump operates smoothly, then attach it to the crankcase. **Note:** *Make sure the O-ring is in place (photo) before installing the pump.*

Installation

5   Hold the pump and metal baffle in place on the crankcase, then install and tighten the mounting screws evenly and securely. After the screws are tight, turn the pump shaft and check for binding, then slide

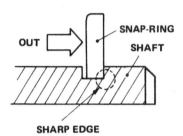

Fig. 2.13   Correct installation of snap-ring that holds the gear on the end of the oil pump shaft (Sec 18)

the nylon gear onto the shaft and mesh it with the steel drive gear on the crankshaft. The flat on the nylon gear must mate with the flat on the oil pump shaft.
6   Install the snap-ring in the groove on the oil pump shaft and make sure it is properly seated. In this case, the sharp edge of the snap-ring should face out.
7   Mounted inside the right crankcase cover is an oil pressure relief valve (photo). It does not require maintenance unless it appears to be clogged with sludge. In that case, remove the cotter key and disassemble the valve, clean the parts thoroughly with solvent and inspect them for damage and wear. Reassemble the valve and install the cotter key (make sure it does not overlap the crankcase cover gasket mating surface).

19   Gearshift mechanism — removal, inspection and installation

1   The gearshift mechanism components are accessible for inspection and repair with the engine in the frame. To begin the disassembly procedure, refer to the appropriate Section and remove the clutch, then remove the shift lever from the shaft (if not already done) and carefully separate the E-clip from the shaft with a screwdriver (photo).
2   Carefully apply pressure to the shift arm (photo) to disengage it from the shift drum, then pull the shaft out of the case (do not lose

18.4   Make sure the O-ring (arrow) is in place before installing the oil pump

18.7   Do not allow the pressure relief valve cotter pin (arrow) to hang up on the gasket surface during installation of the crankcase cover

19.1   Removing the large E-clip with a screwdriver

19.2   Disengaging the shift arm from the drum

the washer on the left end of the shaft). As the shaft is being removed, disengage the stopper arm and spring from the shift drum and
crankcase post and withdraw them with the shaft.

3   Clean all of the parts with solvent and dry them thoroughly.

4   Examine the gearshift mechanism for wear, particularly at the upper arm shift pawls. Make sure the shaft is not bent and check the
springs for cracks and excessive stretch. If the return spring must be
replaced, remove the E-clip and washer. The upper arm must be straight
and free to move at its pivot point (photo).

5   Check the stopper arm, the cam plate and the shift pins for excessive
wear and replace any worn or damaged parts with new ones. The shift
pins and cam plate are attached to the drum by a single TORX screw
(number 30).

6   If the shift shaft oil seal has been leaking, refer to Section 24 and
replace it with a new one. Wrap electrician's tape around the splines
on the end of the shaft to avoid damaging the new seal as the shaft
passes through it.

7   Installation is the reverse of removal. Use a thread locking compound
on the threads and be sure to lubricate the gearshift shaft before inserting it into the crankcase. Engage the return spring on the shift spindle tab and the crankcase post (photo 19.2). Hook the stopper arm
spring under the crankcase projection and be careful to engage the shift
arm with the shift pins.

8   After the components have been reassembled, check the shifter
for proper operation.

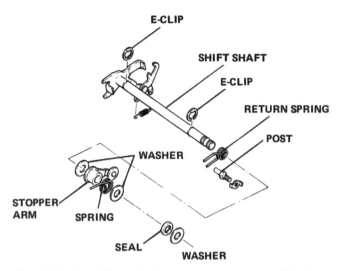

Fig. 2.14 Gearshift mechanism components — exploded view
(Sec 19)

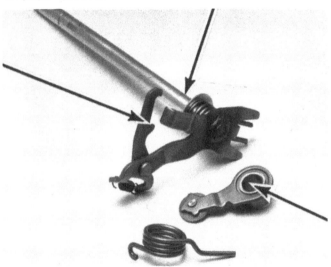

19.4   Check the gearshift components for wear in the areas indicated by the arrows

20.1   Disconnect the wires from the oil pressure sending unit and
the Neutral switch (arrows)

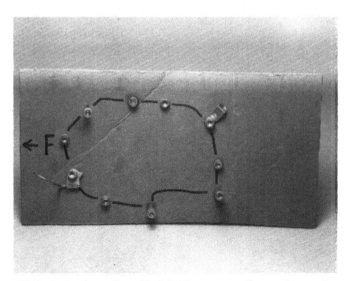

20.2   A simple cardboard bolt holder can save time and prevent
confusion during reassembly

20.3   Slide out the shaft (arrow) while supporting the small idler
gear

20 Alternator/starter clutch — removal and installation

1   The alternator components (rotor and stator) can be removed with the engine in place in the frame, but be sure to separate the shift lever from the shaft and disconnect the wires from the oil pressure sending unit and the neutral switch (photo).

2   Remove the screws and separate the crankcase cover from the engine. It may be a good idea to make a simple holder from a piece of cardboard to ensure that the screws and dowel pins are returned to their original locations when the cover is reinstalled (photo).

3   Pull out on the shaft and remove the small starter idler gear from the crankcase (photo).

4   Remove the bolt from the end of the crankshaft (photo). To keep the crankshaft from turning, attach a chain wrench to the alternator rotor at the extreme edge, next to the crankcase. **Caution:** *Be careful not to damage the pickup coil projections on the outside edge of the rotor.* An alternate method of keeping the crankshaft from turning is to wedge a rolled up rag or a piece of lead between the primary drive gears (on the right side of the engine). If the clutch has been removed, temporarily slip the clutch housing into place on the transmission mainshaft and mesh the primary drive gears.

5   Attach the alternator rotor puller to the rotor. Do not tighten the three puller bolts with a wrench or the starter clutch may be distorted;

tighten them by hand until they bottom lightly. **Caution:** *Do not use a gear puller to try to remove the rotor as damage will result.*

6   Hold the puller and tighten the center bolt (photo) until the rotor separates from the end of the crankshaft. It may be necessary to strike the end of the center puller bolt with a large hammer to dislodge the rotor. If so, one heavy blow is preferred to a series of light ones. As the rotor is removed, the rollers, springs and spring caps of the starter clutch may fall out; be prepared to catch them.

7   If the engine is being disassembled completely, carefully pry the Woodruff key out of the keyway in the crankshat (photo). If no further disassembly is planned, the key can remain in place.

8   Carefully remove the large starter idler gear from the crankshaft (photo).

9   Before installing the alternator, check the starter clutch rollers for wear and flat spots and make sure they move in and out smoothly (photo). Check the hex (Allen) head bolts to make sure they are tight. Do not disassemble the starter clutch unless wear or damage is evident. Check the large idler gear teeth and the bearing and roller contact surfaces for wear and damage. If it is worn or damaged, replace it with a new one.

10   Lubricate the large starter idler gear bearing and slip it onto the crankshaft, then install the Woodruff key in the keyway (if it was removed).

11   Make sure the tapered portion of the crankshaft and the inside of

20.4   Loosening the rotor bolt (note the chain wrench used to keep the rotor from turning)

20.6   Removing the rotor with the special puller

20.7   Prying out the Woodruff key with a small screwdriver

20.8   Slide the large idler gear (arrow) off the crankshaft

the rotor are perfectly clean, then carefully slide the rotor over the end of the crankshaft while lining up the key and slot. Rotate the large starter idler gear clockwise while sliding the rotor into place (this will engage the idler gear and starter clutch rollers properly).

12   Slip the large washer over the bolt (chamfered side out) (photo), then apply thread locking compound to the threads and install the bolt in the end of the crankshaft. Hold the rotor and tighten the bolt to the specified torque.

13   The remainder of the installation procedure is basically the reverse of removal.

---

21   Camshaft drive mechanism — removal, inspection and installation

Front cylinder (No. 1 — right side)

1   The front cylinder camshaft drive mechanism is accessible with the engine in the frame, but if problems occur in the front cylinder camshaft drive, the rear cylinder drive mechanism should be checked as well (which requires removal of the engine). Therefore, it is recommended that the engine be removed from the frame when camshaft drive components require servicing or replacement.

Removal

2   Refer to the appropriate Sections and remove the cylinder head,

cylinder and right side crankcase cover.

3   If the clutch has been removed in the process of completely disassembling the engine, slip the clutch housing onto the transmission mainshaft and mesh the primary drive gears.

4   Wedge a rolled up rag (or a piece of lead) between the primary drive gears, then bend back the lock plate tab (photo).

5   Loosen and remove the large nut on the end of the crankshaft, then slide off the lock plate, the shim, the oil pump drive gear, the primary drive gear, the camshaft drive gear and the square key. Lay the gears out in the correct order to simplify the installation procedure.

6   Remove the hex (Allen) head screw, then support the timing gear and chain while pulling straight out on the timing gear shaft (photo). The timing gear and chain can be withdrawn through the top of the crankcase.

7   Remove the two bolts and separate the cam chain tensioner slipper from the crankcase (photo). If the engine is being completely disassembled, proceed to paragraph 18 and remove the rear cylinder camshaft drive mechanism.

Inspection

8   Clean the parts with solvent and dry them thoroughly, then check the gear and sprocket teeth for wear and damage. Check the timing gear bushing and the timing gear shaft for evidence of seizure, wear and damage. Check the square key and the keyway in the crankshaft for evidence of distortion.

20.9   Check the starter clutch rollers for wear and free movement

20.12   The washer must be installed with the chamfered side (arrow) out

21.4   A rag wedged between the gears will keep them from turning as the large nut is loosened

21.6   The front cylinder timing gear shaft is held in place by a hex head bolt (arrow)

9   Check the cam chain guides and tensioner slipper surfaces for wear (photo). If the rubber is worn or separating from the metal backing plates, replace the parts with new ones.

Installation

10   Install the tensioner slipper and tighten the bolts to the specified torque (use thread locking compound on the bolt threads).
11   Slide the camshaft drive gear (28 teeth) onto the crankshaft, align the keyways in the gear and shaft and slip the square key into place. The notched end of the key should face *in*.
12   Turn the crankshaft until the keyway is pointing directly at the center of the timing gear shaft hole (the crankshaft centerline, the keyway and the timing gear shaft hole centerline should be perfectly aligned as shown in the accompanying illustration).
13   Mesh the chain with the drive sprocket, then lower the chain and timing gear assembly into place in the crankcase. *Note that the timing gear for the front cylinder is stamped with a 1.*
14   Insert a tapered punch into the hole in the timing gear and move the outer stamped gear in relation to the inner gear until the teeth of both gears engage with the teeth of the drive gear far enough to allow the shaft to be installed. **Note:** *The punch mark on the timing gear must be directly opposite the keyway as shown in the accompanying photo.*
15   Lubricate the shaft with clean engine oil or assembly lube, then insert it into the hole in the crankcase and through the timing gear.

21.7   The chain tensioner slipper is attached to the crankcase with two bolts (arrows)

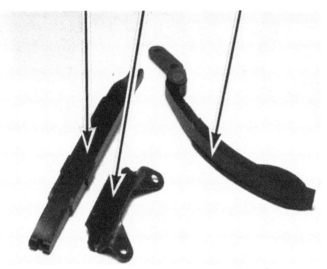

21.9   Check the chain guides and tensioner slipper for wear and damage on the chain contact surfaces (arrows)

21.14   Installing the front cylinder timing gear (note the punch used to move the outer gear relative to the inner gear)

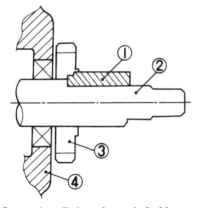

Fig. 2.15   Correct installation of camshaft drive gear and square key on *right* end of crankshaft *(front cylinder)* (Sec 21)

1   Square key   3   Crankshaft drive gear
2   Crankshaft   4   Crankcase

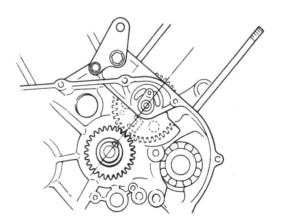

Fig. 2.16   Correct alignment of crankshaft centerline, keyway and timing gear shaft centerline when installing the *front cylinder* timing gear (Sec 21)

21.16   Correct installed position of the key, gears and shim on the right end of the crankshaft

21.19   Remove the hex head bolt and loosen the banjo fitting bolt completely (arrows) before sliding out the rear cylinder timing gear shaft and oil line

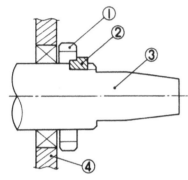

Fig. 2.17   Correct installation of camshaft drive gear and square key on *left* end of crankshaft *(rear cylinder)* (Sec 21)

1   Camshaft drive gear   3   Crankshaft
2   Square key   4   Crankcase

Install the hex head screw and tighten it to the specified torque.
16   Slide the primary drive gear (33 teeth), the oil pump drive gear (23 teeth), the shim (photo) and the lock plate onto the crankshaft. Make sure the tab on the lock plate is engaged in the slot in the shim, then install the nut and tighten it to the specified torque.
17   Using a large pliers, bend up a portion of the lock plate (not the same portion that was flattened to remove the nut) to keep the nut from loosening.

Rear cylinder (No.2 — left side)
Removal
18   Refer to the appropriate Sections and remove the cylinder head, cylinder, alternator and starter idler gears.
19   Remove the hex (Allen) head screw and loosen the oil line banjo fitting bolt (photo). Pull out on the timing gear shaft while unscrewing the bolt, then remove the timing gear shaft, oil line, banjo fitting bolt and copper washers as an assembly (discard the washers and use new ones during installation).
20   Withdraw the timing gear and chain assembly through the top of the crankcase, then remove the bolts and separate the tensioner slipper from the crankcase.
21   Flatten the lock tab, then loosen, but do not remove, the large nut on the left end of the crankshaft. Wedge a rolled up rag (or piece of lead) between the balancer drive gears to keep the crankshaft from turning (photo).
22   *If the engine is being completely disassembled,* flatten the lock tab and loosen the nut on the balancer shaft (photo), then remove the nuts, lock plates, gears and keys and proceed to Section 23. *If the engine is not being completely disassembled,* remove the nut, lock plate, gears and key from the crankshaft only.

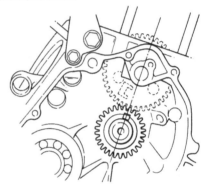

Fig. 2.18   Correct alignment of crankshaft centerline, keyway and timing gear shaft centerline when installing the *rear cylinder* timing gear (Sec 21)

Inspection
23   Refer to paragraphs 8 and 9 for the inspection procedures.

Installation
24   Install the tensioner slipper and tighten the bolts to the specified torque (use thread locking compound on the bolt threads).
25   Slide the camshaft drive gear (28 teeth) onto the crankshaft, align the keyways in the gear and shaft and slip the square key into place. The notched end of the key should face *in.*
26   Align the crankshaft and keyway as described in paragraph 12, then install the timing gear and shaft as outlined in paragraphs 13 and 14. *Note that the rear cylinder timing gear is stamped with a 2 rather than a 1.* Be sure that the punch mark on the gear is correctly aligned with the keyway as described in paragraph 14 before installing the timing gear shaft (photo).
27   Lubricate the shaft with clean engine oil or assembly lube, then insert it, along with the oil line and banjo fitting bolt (be sure the new copper washers are in place on each side of the banjo fitting before the bolt is installed).
28   Carefully thread the banjo fitting bolt and the hex screw into place, then tighten them both to the specified torque.
29   Refer to Section 22 and install the balancer driven gear (if it was removed).
30   Position the balancer drive gear over the end of the crankshaft with the punch mark (dished side) facing out. Align the keyway in the gear with the key and turn the balancer shaft until the punch marks in the gears are opposite each other, then slide the gear onto the crankshaft (photo) and install the lock plate and the nut.
31   Tighten the crankshaft nut to the specified torque, then bend up the unused tab of the lock plate to keep the nut from loosening (if both tabs on the lock plate have been used, replace the plate with a new one).

21.21   Loosening the large nut on the left side of the crankshaft (note the rag wedged between the balancer shaft gears)

21.22   Loosening the balancer shaft nut

21.26   Installing the rear cylinder timing gear (note the punch mark on the gear aligned with the key in the crankshaft)

21.30   The punch marks (arrows) must be aligned when installing the balancer shaft drive and driven gears

22.3a   Install the thrust washer, . . .

22.3b   . . . the inner plate and key, . . .

22.4   . . . the balancer gear assembly, . . .

22   Balancer driven gear — removal, inspection and installation

1   Refer to Section 21, paragraph 21 for the procedure to follow when separating the balancer driven gear from the shaft. Note that the drive gear on the crankshaft must be in place in order to loosen the large nut on the end of the balancer shaft. **Note:** *Do not separate the driven gear from the hub unless new parts are required*. If the gear and hub are separated, align the punch marks when they are reassembled.

2   Check the gear teeth for wear and damage and inspect the keyways in the hub and balancer shaft for distortion. If the springs appear to have sagged or if they are broken or cracked, replace the springs and pins with new parts (the pins are used only in every other spring). If both tabs of the lock plate have been bent, replace the lock plate with a new one when reinstalling the gear.

3   Begin installation by slipping the thrust washer over the shaft, followed by the square key and the inner plate (photos).

4   Align the keyway with the key and carefully slide the gear onto the shaft with the punch marks facing out (photo). **Caution:** *Do not push on the gear — push only on the hub or the parts may separate.*

5   Install the outer plate, the lock plate and the nut (photo). Note that the tab on the lock plate must engage with the keyway in the shaft and the nut must be installed with the relieved side facing *in*.

6   The nut can be tightened to the specified torque only after wedg-

22.5   . . . the outer plate and the lock plate and nut, then install the drive gear, tighten the nut and bend up the unused lock plate tab (arrow)

23.3   *Left* side crankcase bolt locations

23.4a   *Right* side crankcase bolt locations

23.4b   Position the shift drum so the cam plate will pass through the cutouts in the crankcase

ing a rolled up rag (or piece of lead) between the gears (refer to Section 21).

7 After the nut has been tightened, bend up the tab on the lock plate to keep it from loosening.

23 Crankcase — disassembly and reassembly

1 To examine and repair or replace the crankshaft, connecting rods, bearings, transmission components and balancer shaft, the crankcase must be split into two parts.

2 Before this can be done, the clutch (Section 13), the oil pump (Section 16), the gearshift mechanism (Section 19), the alternator (Section 20), the cam drive components (Section 21), and the balancer shaft gears must be removed. **Note:** *Do not separate the shaft drive output gearbox from the engine. If gearbox service or repair work is required, take the engine to a Yamaha dealer service department.*

Disassembly

3 Remove the bolts from the left side of the crankcase (photo), then support the engine on blocks of wood or a homemade engine stand with the *right* side facing up. To prevent distortion of the cases, loosen the bolts in three steps, following a criss-cross pattern. It may be helpful to make a simple holder from a piece of cardboard to ensure that the bolts are returned to their original locations when the crankcase is reassembled.

4 Remove the bolts from the right side (photo), then align the shift drum cam plate with the cutouts in the right crankcase half (photo).

5 Gently tap the upper (right) case with a soft-faced hammer to break the seal, then carefully lift it away from the lower (left) case half. If necessary use a screwdriver or pry bar to help separate the crankcase halves. **Caution:** *Do not pry between the case sealing surfaces to separate them. Pry only at the slots provided in front of the number one (1) cylinder and along the bottom of the cases (photo).* If resistance is encountered, double check to make sure that all of the bolts have been removed. Once the crankcase has been separated, remove the dowel pins from the bolt holes.

6 The balancer shaft, crankshaft and connecting rods, shift drum and transmission shafts will remain in place in the left crankcase half. Lift out the balancer shaft and the crankshaft and store them where they will not be damaged by moisture or contact with other components. Refer to the following Sections (24 through 26) for transmission removal and servicing procedures for crankcase-mounted components (bearings, seals, clutch actuating mechanism, etc.).

Reassembly

7 Before rejoining the cases, the transmission components, the crankshaft and connecting rods and the balancer shaft must be in place in the left crankcase section. **Note:** *The balancer shaft must be install-*

ed with the threaded end in the left crankcase. The left crankcase should be supported on wood blocks or a homemade engine stand.

8 Clean the mating surfaces with lacquer thinner or acetone, then apply a thin coat of Yamabond 4 semi-drying liquid gasket to *both* crankcase mating surfaces (be sure to follow the directions on the container).

9 Position the left-hand connecting rod in the front cylinder opening and the right-hand rod in the rear cylinder opening, then install the three dowel pins in the left crankcase.

10 Make sure the transmission is in Neutral, then slowly and carefully lower the right crankcase section into position. The shift drum cam plate should pass through the cutouts in the crankcase opening.

11 Make sure the cases are seated, then install the right side bolts. Turn the engine over and install the left side bolts. Tighten the bolts gradually, in the sequence shown in the accompanying illustration, until the specified torque is reached. Note that the large bolt has a different torque specification than the smaller bolts.

12 Turn the crankshaft and transmission shafts several times by hand and check for any obvious binding or rough spots. *Do not proceed with the engine reassembly if the crankshaft and transmission shafts do not rotate freely.*

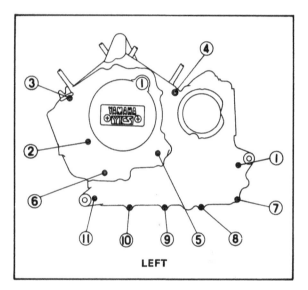

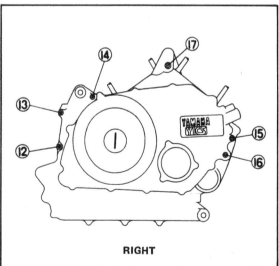

Fig. 2.19 Crankcase bolt *tightening* sequence (Sec 23)

23.5 When separating the crankcases, pry only at the designated areas

24 Crankcase components — inspection and servicing

1   After the crankcases have been separated and the crankshaft, balancer shaft, shift drum and forks and transmission components removed, the crankcases should be cleaned thoroughly with new solvent and dried with compressed air. **Note:** *Do not submerge the left crankcase section in solvent as the lubricant in the output gearbox will be washed out.* All oil passages should be blown out with compressed air and all traces of old gasket sealant should be removed from the mating surfaces. **Caution:** *Be very careful not to nick or gouge the*

crankcase mating surfaces or leaks will result. Check both crankcase sections very carefully for cracks and other damage.

2   Carefully pry the shift shaft seal out of the left crankcase with a large screwdriver (photo).

3   Apply a thin layer of multi-purpose grease to the outer edge of the new seal and press it into position with a suitably sized socket (photo). *The manufacturer's marks or numbers must face out.*

4   Remove the screw from the left crankcase and withdraw the clutch actuating shaft (photo). Check the roller bearing for wear (photo). If it is worn or damaged (which is not likely), remove the snap-ring and carefully pry out the seal with a large screwdriver. When the seal is

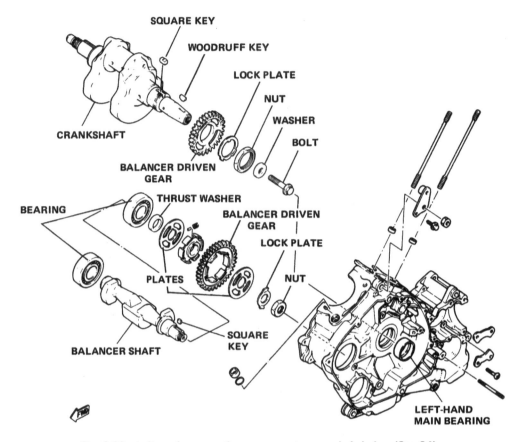

Fig. 2.20 *Left* crankcase section components — exploded view (Sec 24)

24.2   Removing the shift shaft seal with a large screwdriver (be very careful not to damage the crankcase)

24.3   Installing the new shift shaft seal with a socket

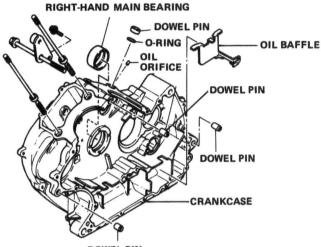

RIGHT-HAND MAIN BEARING

DOWEL PIN

O-RING

OIL BAFFLE

OIL ORIFICE

DOWEL PIN

DOWEL PIN

CRANKCASE

DOWEL PIN

Fig. 2.21 *Right* crankcase section components — exploded view
(Sec 24)

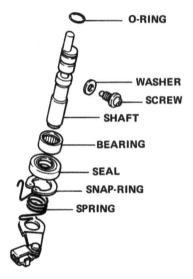

O-RING

WASHER

SCREW

SHAFT

BEARING

SEAL

SNAP-RING

SPRING

Fig. 2.22 Clutch actuating shaft components — exploded view
(Sec 24)

removed, the bearing will slide out of place. Lubricate and install the new bearing, then apply a thin layer of multi-purpose grease to the outer edge of the new seal and press it into position with a suitably sized socket. *The manufacturer's marks or numbers must face out.* Install the snap-ring with the sharp edge facing out.

5   Check the shaft and O-ring for wear as well (photo), then lubricate the O-ring, fill the groove with grease and reinstall the shaft in the crankcase. Install the screw (and washer) and tighten it securely.

6   Rotate all of the ball bearings with your fingers (one at a time, of course) and check for noise, binding, rough spots and excessive play. If wear or damage is evident, replace the bearing(s) with a new one. **Note:** *The bearings should be removed only after the case has been heated in an oven to a temperature of from 205 to 257°F (95 to 125°C).* Never use a torch as the crankcase may be distorted and ruined. In the case of a bearing in the left crankcase section, have the removal and installation done by a Yamaha dealer (since the output gearbox assembly must be disassembled first — a procedure which is beyond the scope of the home mechanic).

7   If an oven is not available, the bearings can be driven out with a hammer and punch (photo), but damage to the crankcase could occur if this method is used. When installing the new bearings, apply pressure only to the *outer* race or the bearing will be damaged. *The bearings must be installed with the manufacturer's marks or numbers facing out.*

24.4a   Remove the screw (arrow) to separate the clutch actuating shaft from the crankcase

24.4b   The roller bearing (arrow) is held in place by the snap-ring and seal

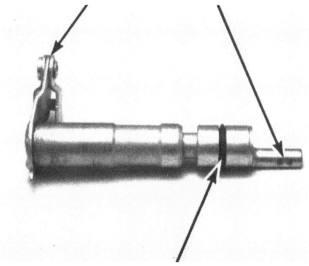

24.5   Check the cam, O-ring and lever pivot (arrows) on the clutch acutating shaft for wear and damage

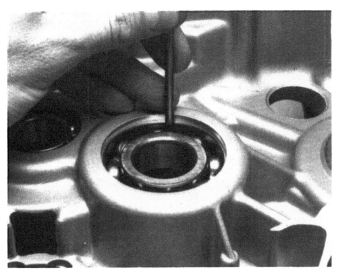

24.7   Removing a bearing by tapping on the outer race

25   Crankshaft — removal and installation

1   Crankshaft removal is a simple matter of lifting it out of place once the crankcase has been separated.
2   Before installing the crankshaft, the connecting rods must be attached and the main bearings must be inspected and replaced as necessary.
3   When installing the crankshaft, lubricate the main bearing journals with clean engine oil, assembly lube or molybdenum disulfide grease and carefully slide the tapered end of the shaft into the main bearing in the *left* crankcase section. Be careful not to nick or gouge the bearing with the end of the shaft.
4   Turn the crankshaft by hand to check for any obvious binding, then position the connecting rods so the left-hand rod is in the front cylinder opening and the right-hand rod is in the rear cylinder opening.

26   Shift drum/shift forks — removal, inspection and installation

Removal

1   Support the shift forks and pull straight up on the shift fork shafts, then separate the forks from the gears.
2   Pull straight up on the shift drum to remove it from the crankcase.

26.3   Check the edges of the grooves and the shift pins in the drum (arrows) for wear and damage

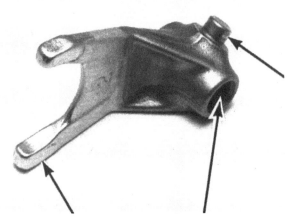

26.5   Check the shift fork ends, the bore and the guide pin (arrows) for wear

26.8   Each shift fork is stamped with a number (which must face the *left* side of the crankcase)

26.9   Correct installed position of the shift forks (1 through 3), the short shaft (4) and the long shaft (5)

Inspection

3   Check the edges of the grooves in the drum for signs of excessive wear. Inspect the end plate, the shift pins and the cam plate for wear (photo) and make sure the TORX screw is tight. If wear is evident, new parts will be required.

4   Check the ball bearing for smooth operation. If noise or binding is evident, the bearing must be replaced with a new one. To replace it, remove the TORX screw from the end of the drum, slip off the end plate, the shift pins and the cam plate, then remove the bearing and install a new one. *The numbers on the bearing should face out when it is installed.*

5   Check the shift forks for distortion and wear, especially at the fork ends. If they are discolored or severely worn they are probably bent. If damage or wear is evident, check the shift fork groove in the corresponding gear as well. Inspect the guide pins and the shaft bore for excessive wear and distortion and replace any defective parts with new ones (photo).

6   Check the shift fork shafts for evidence of wear, galling and other damage. Make sure the shift forks move smoothly on the shafts. If the shafts are worn or bent, replace them with new ones.

Installation

7   Lubricate the grooves in the shift drum (or the guide pins on the forks) with molybdenum disulfide grease, then install the drum in the left crankcase half.

8   Each shift fork is marked with a number (photo) and must be installed with the number toward the *left* side of the crankcase.

9   Lubricate the fork ends, then engage shift fork number 2 in the mainshaft third (3rd) gear groove and the center shift drum groove. Lubricate the short shaft, then slide it through the shift fork bore and into the recess in the crankcase (photo).

10   Lubricate the fork ends, then engage shift fork number 1 in the countershaft fifth (5th) gear groove and the lower shift drum groove. Engage shift fork number 3 in the countershaft fourth (4th) gear groove and the upper shift drum groove, then lubricate the remaining shaft and slide it through both shift fork bores and into the recess in the crankcase.

11   Turn the shift drum (and transmission gears) as necessary to place the transmission in the Neutral position, then make sure the gears turn freely before proceeding with the engine reassembly procedures.

---

27   Transmission shafts — disassembly, inspection and reassembly

1   **Note:** *In order to disassemble the countershaft gears, a snap-ring pliers is an absolute necessity. The mainshaft cannot be disassembled and reassembled without an hydraulic press. As a result, if the gears on the mainshaft (or the mainshaft itself) require replacement, we recommend that it be done by a Yamaha dealer service department.*

Disassembly

2   Remove the snap-ring from the end of the countershaft (photo), then lift off the thrust washer and the countershaft first (1st) gear.

3   Slide off the countershaft fourth (4th) gear, then remove the next snap-ring (photo). Remove the thrust washer, then slide off the countershaft third (3rd) gear while lifting out the mainshaft assembly.

4   Slide off the countershaft fifth (5th) gear, then remove the remaining snap-ring and thrust washer (photo) and slide off the countershaft second (2nd) gear. The countershaft cannot be removed without disassembling the shaft drive output gearbox (a procedure that must be done by a Yamaha dealer service department).

5   Lay the parts out in the order of disassembly, then clean each part, one at a time, with solvent and dry them thoroughly. Make sure the oil holes are clear.

Inspection
Countershaft

6   Check the gear teeth, the gear dogs and the shift fork grooves for cracks, heat discoloration and excessive wear. If the gear dogs are rounded off, replace the gears with new ones and check the slots in the gears that mate with the dogs.

7   Check the splines on the shaft and the splines in the gears for wear and distortion. Make sure each of the countershaft gears slides freely on the shaft splines (photo).

27.2   Removing the snap-ring from the end of the countershaft

27.3   Removing the snap-ring that attaches the countershaft third (3rd) gear to the shaft

27.4   The remaining snap-ring (arrow) retains the countershaft second (2nd) gear

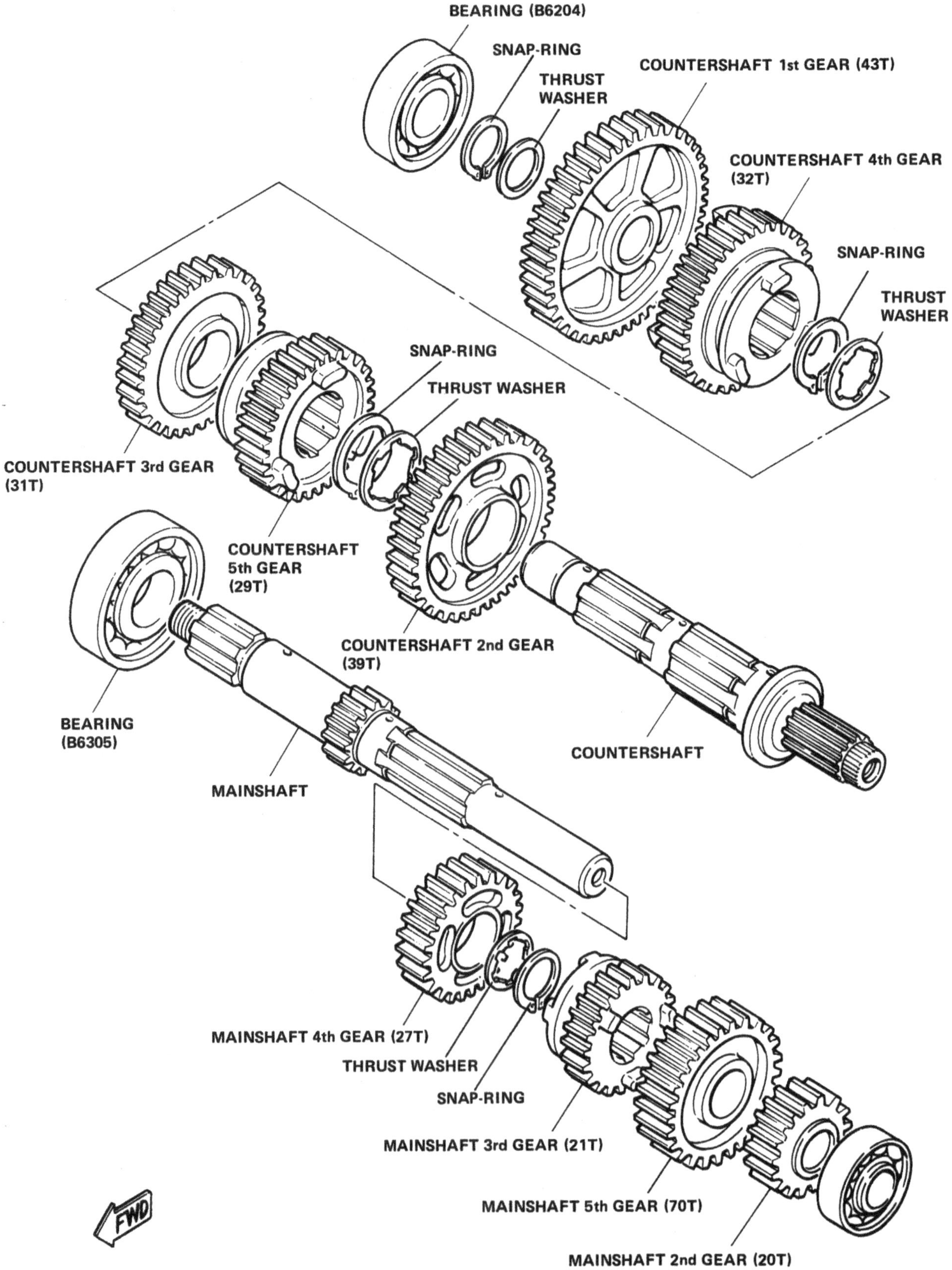

84

BEARING (B6204)

SNAP-RING

THRUST WASHER

COUNTERSHAFT 1st GEAR (43T)

COUNTERSHAFT 4th GEAR (32T)

SNAP-RING

THRUST WASHER

SNAP-RING

THRUST WASHER

COUNTERSHAFT 3rd GEAR (31T)

COUNTERSHAFT 5th GEAR (29T)

COUNTERSHAFT 2nd GEAR (39T)

COUNTERSHAFT

BEARING (B6305)

MAINSHAFT

MAINSHAFT 4th GEAR (27T)

THRUST WASHER

SNAP-RING

MAINSHAFT 3rd GEAR (21T)

MAINSHAFT 5th GEAR (70T)

MAINSHAFT 2nd GEAR (20T)

FWD

Fig. 2.23 Transmission shaft components — exploded view (Sec 27)

27.7 Inspect the gear teeth, the splines, the shift grooves and the gear dogs (arrows) for signs of wear and damage

27.8 Check the dog slots and the bearings (arrows) for wear

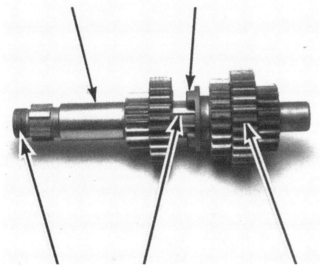

27.10a Inspect the mainshaft components for wear and damage in the areas indicated by the arrows

27.10b Check the mainshaft gear shift grooves and dog slots (arrows) for wear

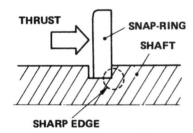

Fig. 2.24 When installing the transmission shaft snap-rings, the sharp edge must face *away* from the thrust washer (Sec 27)

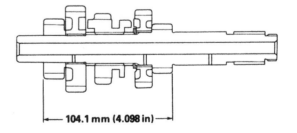

Fig. 2.25 Mainshaft gear cluster width (Sec 27)

8 Check the gear bushings (photo) and the corresponding shaft journals for wear, galling and heat discoloration. Make sure they rotate freely on the shaft without excessive play. If a gear requires replacement due to wear or damage, have the gear it mates against on the mainshaft replaced also.

9 Check the thrust washers and snap-rings for wear and distortion. Replace any worn or damaged parts with new ones (if in doubt as to the condition of the snap-rings, replace them with new ones — the cost is minimal and it may prevent severe damage that could occur if a snap-ring breaks or shifts during engine operation).

Mainshaft

10 Check the mainshaft gears and shaft in the same manner as the countershaft components (photos). Note that if mainshaft components require replacement, it should be done by a Yamaha dealer service department. If a gear requires replacement due to wear or damage, replace the gear that it mates against on the countershaft as well.

Reassembly

11   Reassemble and install the components in the reverse order of disassembly. Use the exploded view as a guide. As the parts are assembled, lubricate the contact surfaces with molybdenum disulfide grease. Also, make sure the snap-rings are securely seated in their grooves with the sharp edge on each snap-ring facing out.

12   Before installing the mainshaft, measure the width of the gear cluster with a dial or Vernier caliper. If it is not the width shown on the accompanying illustration, have the mainshaft second gear position changed by a Yamaha dealer service department (an hydraulic press is required to move it).

---

28   Main and connecting rod bearing inspection — general note

---

1   Even though main and connecting rod bearings are generally replaced with new ones during the engine overhaul, the old bearings should be retained for close examination as they may reveal valuable information about the condition of the engine.

2   Bearing failure occurs mainly because of lack of lubrication, the presence of dirt or other foreign particles, overloading the engine and/or corrosion. Regardless of the cause of bearing failure, it must be corrected before the engine is reassembled to prevent it from happening again.

3   When examining the bearings, remove the rod bearings from the connecting rods and the rod caps and lay them out on a clean surface in the same general position as their location on the crankshaft journal. This will enable you to match any noted bearing problems with the corresponding side of the crankshaft journal.

4   Dirt and other foreign particles get into the engine in a variety of ways. It may be left in the engine during assembly or it may pass through filters or breathers. It may get into the oil and from there into the bearings. Metal chips from machining operations and normal engine wear are often present. Abrasives are sometimes left in engine components after reconditioning operations such as cylinder honing, especially when parts are not thoroughly cleaned using the proper cleaning methods. Whatever the source, these foreign objects often end up imbedded in the soft bearing material and are easily recognized. Large particles will not imbed in the bearing and will score or gouge the bearing and journal. The best prevention for this cause of bearing failure is to clean all parts thoroughly and keep everything spotlessly clean during engine reassembly. Frequent and regular oil and filter changes are also recommended.

5   Lack of lubrication or lubrication breakdown has a number of interrelated causes. Excessive heat (which thins the oil), overloading (which squeezes the oil from the bearing face) and oil leakage or throw off (from excessive bearing clearances, worn oil pump or high engine speeds) all contribute to lubrication breakdown. Blocked oil passages will also starve a bearing and destroy it. When lack of lubrication is the cause of bearing failure, the bearing material is wiped or extruded

from the steel backing of the bearing. Temperatures may increase to the point where the steel backing and the journal turn blue from overheating.

6   Riding habits can have a definite effect on bearing life. Full throttle low speed operation, or lugging the engine, puts very high loads on bearings, which tend to squeeze out the oil film. These loads cause the bearings to flex, which produces fine cracks in the bearing face (fatigue failure). Eventually the bearing material will loosen in pieces and tear away from the steel backing. Short trip driving leads to corrosion of bearings, as insufficient engine heat is produced to drive off the condensed water and corrosive gases produced. These products collect in the engine oil forming acid and sludge. As the oil is carried to the engine bearings, the acid attacks and corrodes the bearing material.

7   Incorrect bearing installation during engine assembly will lead to bearing failure as well. Tight fitting bearings which leave insufficient bearing oil clearances result in oil starvation. Dirt or foreign particles trapped behind a bearing insert result in high spots on the bearing which lead to failure.

8   To avoid bearing problems, clean all parts thoroughly before reassembly, double check all bearing clearance measurements and lubricate the new bearings with engine assembly lube or molybdenum disulfide grease during installation.

---

29   Connecting rods and bearings — removal, inspection and installation

---

1   Before removing the connecting rods from the crankshaft, check the side clearance with a feeler gauge (photo). If the side clearance is greater than specified, new connecting rods will be required for engine reassembly.

Removal

2   Using a center punch and hammer, carefully mark the crankshaft, the connecting rods and the caps so they can be reinstalled in the same position on the same side of the crankshaft journal. Loosen the cap nuts on one connecting rod in three steps, carefully lift off the cap and bearing insert, then carefully remove the rod and remaining bearing insert from the crankshaft journal. Temporarily reassemble the rod, the bearing and the cap to prevent mixing up parts.

3   Repeat the procedure for the remaining connecting rod. Be very careful not to nick or scratch the crankshaft journal with the rod bolts.

4   Without mixing them up, clean the parts with solvent and dry them thoroughly. Make sure the oil holes are clear.

Inspection

5   Check the connecting rods for cracks and other obvious damage. Lubricate the piston pin, install it in the rod and check for play (photo). If the pin is loose, replace the connecting rod and/or the pin.

29.1   Checking the connecting rod side clearance with a feeler gauge

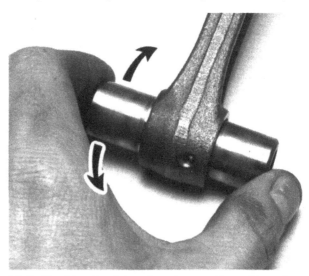

29.5   Checking the piston pin and connecting rod bore for wear

6   Refer to Section 28 and examine the connecting rod big end bearing inserts. If they are scored, badly scuffed or appear to have been seized, new bearings must be installed. Always replace the bearings in both connecting rods as a set. If they are badly damaged, inspect the corresponding side of the crankshaft journal. Evidence of extreme heat, such as discoloration, indicates that lubrication failure has occurred. Be sure to thoroughly check the oil pump and oil pressure relief valve as well as all oil holes and passages before reassembling the engine.
7   If the bearings and journals appear to be in good condition, check the oil clearances as follows:
8   Remove the connecting rod caps from the rods. Remove the bearing inserts and wipe the bearing surfaces of the rods and caps with a clean, lint-free cloth. *They must be spotlessly clean.*
9   Clean the back sides of the upper bearing inserts, then lay them in place in the connecting rods. Make sure that the tab on the bearing fits into the recess in the rod. Also, the holes in the rod and bearing insert must line up. Do not hammer the bearing insert into place and be very careful not to nick or gouge the bearing surface. *Do not lubricate the bearing at this time.*
10   Clean the back sides of the other bearing inserts and install them in the rod caps. Again, make sure the tab on the bearing fits into the recess in the cap and do not apply any lubricant. It is critically important to ensure that the mating surfaces of the bearings and connecting rods are perfectly clean and oil free when they are assembled.
11   Hold the connecting rods in a vise equipped with soft jaws, then clean the crankshaft journal thoroughly and lay the crankshaft in place in the rods. Make sure the rods are positioned exactly as they were before removal from the crankshaft (the Y on the rods must face the tapered end of the crankshaft) (photo) and do not apply excessive pressure with the vise. Also, be extremely careful not to drop the crankshaft; make sure the rods can support its weight.
12   Trim two pieces of the appropriate type Plastigage so that they are slightly shorter than the width of the connecting rod bearings and lay them in place on the rod journal, parallel with the journal axis (photo).
13   Clean the connecting rod cap bearing faces and gently install the rod caps in place. Make sure the mating marks on the caps are on the same side as the marks on the connecting rods. Lubricate the threads with molybdenum disulfide grease, then install the nuts and tighten them to the specified torque, working up to it in three steps. **Caution:** *When tightening the rod cap nuts, apply continuous torque between 24 and 27 ft-lbs (34 and 38 Nm). If tightening is interrupted, loosen the nuts to less than 24 ft-lb (34 Nm) and start again. Do not move the crankshaft at any time during this operation.*
14   Remove the rod caps, being very careful not to disturb the Plastigage. Compare the width of the crushed Plastigage to the scale printed on the Plastigage container (photo) to obtain the oil clearance. Compare the results to the Specifications to make sure the clearance is correct. If the clearance is not correct, make sure that no dirt or oil was between the bearing inserts and the connecting rods or caps when

the clearance was measured. Remove all traces of the Plastigage with a plastic or wooden scraper, then wipe the crankshaft journal and bearing insert with a solvent-soaked cloth.
15   As a general rule, bearings are replaced as a matter of course whenever the engine is disassembled, but if the clearances are within the specified limits, the original bearings can be reinstalled when the engine is assembled. However, it should be noted that if the clearances are within the specified limits, but very close to the high side of the limits, it would be wise to go ahead and replace the bearings with new ones, since the engine is already apart.
16   If the oil clearances are excessive, or if the bearings are damaged, refer to Section 31 to determine the correct bearings to use during reassembly. Also, refer to the appropriate Section and inspect the crankshaft journals and main bearings before proceeding.

Installation
17   Once the crankshaft has been cleaned and inspected and the decision has been made concerning bearing replacement, the connecting rods can be reinstalled on the crankshaft. **Note:** *If new bearings are being used, check the oil clearance as described in paragraphs 11 through 14 before final installation of the connecting rods. If the clearances are within the specified limits, proceed with the installation. Never assume that the clearances are correct even though new bearings are involved.*

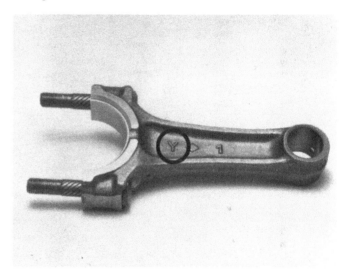

29.11   The connecting rods must be installed with the Y mark (circled) facing the tapered (alternator) end of the crankshaft

29.12   Lay the Plastigage strips on the bearing journal, parallel to the journal axis

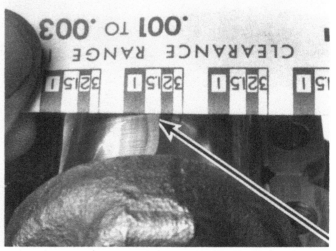

29.14   The oil clearance is obtained by comparing the crushed Plastigage to the scale printed on the container (in this example the clearance is approximately 0.0015 in)

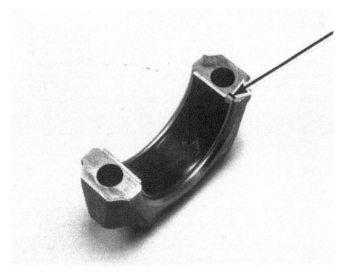

29.18   The tab on the bearing (arrow) must fit into the recess in the rod or cap (note the lubricant on the bearing face)

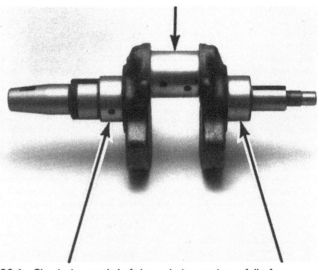

30.1   Check the crankshaft journals (arrows) carefully for wear and damage and make sure the oil passages are clean and clear

30.2   Measuring a main bearing journal diameter with a micrometer

30.7a   Using a telescoping gauge to determine the diameter of the main bearing

30.7b   Measuring the telescoping gauge with a micrometer to obtain the actual size in inches

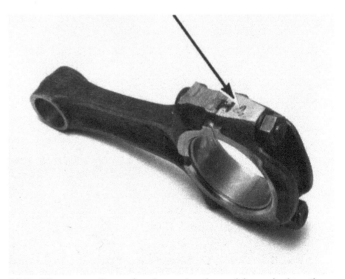

31.2   The connecting rod code number (arrow) is marked on the side of the rod (and/or cap)

18   Make sure the bearing faces are perfectly clean, then apply a uniform layer of clean grease (preferably molybdenum disulfide) to the bearing faces (photo). The tab on the bearing must be engaged in the recess in the cap or rod and the oil holes in the bearings and rods must line up. Lay the crankshaft in place in the rods (make sure the mating marks made during disassembly are matched up correctly).

19   Install the rod caps in their original positions, then lubricate the rod bolt threads with molybdenum disulfide grease and install the nuts. Make sure you install the caps on the correct rod and match up the mating marks made during disassembly.

20   Tighten the nuts to the specified torque. Again, work up to the final torque in three steps. **Caution:** *When tightening the rod cap nuts, apply continuous torque between 24 and 27 ft-lbs (34 to 38 Nm). If tightening is interrupted, loosen the nuts to less than 24 ft-lb (34 Nm) and start again.*

21   After the connecting rods have been installed, rotate them by hand and check for any obvious binding.

22   As a final step, the connecting rod big end side clearances must be rechecked. Slide the connecting rods to one side of the journal and slip a feeler gauge between the side of the connecting rod and the crankshaft throw. Be sure to compare the measured clearance to the Specifications to make sure it is correct.

---

30   Crankshaft and main bearings – cleaning and inspection

Note: *Crankshafts installed in RK model engines have two digits stamped into the left side crank web. The left digit (2) is the crankshaft ID number; the right digit is the crankpin (journal) size code. Crankshafts that do not have the ID number "2" cannot be installed in an RK engine.*

1   Clean the crankshaft with solvent (be sure to clean the oil holes with a stiff brush and flush them with solvent) and dry it thoroughly. Check the main and connecting rod bearing journals for uneven wear, scoring, pitting and metal separation (photo). Check the remainder of the crankshaft for cranks and damage.

2   Using an appropriate size micrometer, measure the diameter of the main (photo) and connecting rod journals and compare the results to the Specifications. By measuring the diameter at a number of points around the journal's circumference you will be able to determine whether or not a journal is worn out-of-round. Take the measurements at each end of the connecting rod journal, near the crankshaft throws, to determine if it is tapered. Record the journal diameter measurements as they will be needed to determine the main bearing oil clearances.

3   If the crankshaft journals are damaged, tapered, out-of-round or worn beyond the limits given in the Specifications, replace it with a new one.

4   Have the crankshaft runout checked by a dealer service department, motorcycle repair shop or automotive machine shop. This is done by laying the main bearing journals in V-blocks and reading the runout at the shaft ends with a dial indicator. If the precision measuring tools required for Paragraph 2 above are not available, the dealer service department or repair shop can also make the journal measurements.

5   Check the security of the ball bearings which are used to plug the oil passages in the crankshaft. They occasionally work loose, causing lubrication problems and subsequent bearing failure. A loose ball can be carefully staked back into place after being treated with a stud and bearing mount liquid or epoxy.

6   Refer to Section 28 and examine the main bearings. If they are scored, badly scuffed, discolored from excessive heat or appear to have been seized, new bearings must be installed (a procedure that must be done by a Yamaha dealer service department). If they are badly damaged, the crankshaft journals are probably damaged as well. If so, a new crankshaft will be required. If lubrication failure has occurred, be sure to thoroughly check the oil pump and oil pressure relief valve as well as all oil holes and passages before reassembling the engine.

7   If the bearings appear to be in good condition, check the main bearing clearances as follows: Using a telescoping gauge (two-inch capacity), determine the size of the main bearing insert in the right crankcase half (photo). Measure the telescoping gauge with a micrometer (photo) and record the measurement. Repeat the check at a point 90~ from the first one, measure the gauge and record the measurement.

8   Repeat the procedure for the main bearing insert in the left crankcase half.

9   Subtract the *largest* left side crankshaft journal outside diameter (recorded in Paragraph 2) from the *smallest* left side main bearing diameter recorded during the measurement procedure. The result is the bearing oil clearance for the left-hand main bearing and crankshaft journal.

10   Repeat the procedure for the right-hand bearing journal.

11   If the oil clearances are greater than the specified limit, take both crankcase sections and the crankshaft to a Yamaha dealer service department so new main bearing inserts can be installed.

12   As a general rule, main bearings are replaced as a matter of course whenever the engine is dismantled, but if the oil clearances are within the specified limits the original bearings can remain in place when the engine is assembled. However, it should be noted that if the clearances are within the specified limits, but very close to the high side of the limits, it would be wise to go ahead and have the bearings replaced with new ones, since the engine is already apart.

---

31   Connecting rod bearings – selection

---

1   New connecting rod bearings are selected on the basis of the condition of the old bearings, the size and condition of the crankshaft journal and the recommended oil clearance. It is done by cross referencing numbered codes stamped into the crankshaft and marked on the connecting rods.

2   The connecting rod inside diameter code number is marked on the side of the rod (photo).

3   The journal outside diameter code number is stamped into the crankshaft counterweight (photo).

4   To determine the correct bearing, find the crankshaft outside diameter number and the connecting rod inside diameter number

31.3   The crankshaft journal size code number is stamped into the crankshaft throw (arrow)

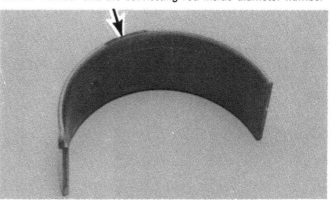

31.4   The bearing insert color code is located on the side of the bearing (arrow)

as described above. Subtract the number on the crankshaft from the number on the connecting rod, then refer to the accompanying chart for the bearing to use. For example, the connecting rod in the photo is marked with a 4 and the crankshaft is stamped with a 1. The bearing that should be used would be color-coded brown (4 - 1 = 3; no. 3 on the chart is a brown coded bearing). The color in the box denotes the bearing required. The bearing inserts will be appropriately color-coded (photo).

5　Repeat the procedure to determine the correct bearing for the remaining connecting rod.

6　During reassembly, be careful not to accidentally interchange the bearings and remember to double check the oil clearances with Plastigage.

32　Crankshaft main bearings — selection

Since the removal and installation of the main bearings must be done by a Yamaha dealer service department (due to the need for special tools and equipment), the bearing selection procedure should be left to them as well. Be sure to take the crankshaft and both crankcase sections to the dealer when new main bearings are required.

33　Initial startup after major repair

Note: *Make sure the cooling system is checked carefully (especially the coolant level) before starting and running the engine.*

1　Make sure the engine oil level is correct, then remove both spark plugs from the engine. Place the engine STOP switch in the Off position.

2　Turn on the key switch and crank the engine over with the electric starter until the oil pressure indicator light goes off (which indicates that oil pressure exists). Replace the spark plugs, hook up the wires and turn the switch to On.

3　Make sure there is fuel in the tank, then turn the petcock to the Prime position and operate the choke.

4　Start the engine and allow it to run at a moderately fast idle until it reaches operating temperature. **Warning:** *If the oil pressure indicator light does not go off, or if it comes on while the engine is running, stop the engine immediately.*

5　Check carefully for oil leaks and make sure the transmission and controls, especially the brakes, function properly before road testing the machine. Refer to Section 34 for recommended break-in procedures.

Bearing color code	
No. 1	Blue
No. 2	Black
No. 3	Brown
No. 4	Green
No. 5	Yellow

Fig. 2.26　Connecting rod bearing selection chart (Sec 31)

6　Upon completion of the road test, and after the engine has cooled down completely, retorque the cylinder head nuts and recheck the valve clearances.

34　Recommended break-in procedure

1　An engine that has had extensive work such a new piston rings, new main and/or connecting rod bearings or new transmission parts must be carefully broken in to realize the maximum possible benefits from the repairs.

2　The break-in procedure allows the new parts to wear in under controlled conditions and conform to the surfaces which they bear against.

3　Generally, the break-in procedure requires that the engine be allowed to spin freely under light loads without over revving it or continuously running it at a constant speed. Do not lug the engine (by applying large throttle openings at low speeds) and do not allow it to idle for long periods of time. These guidelines should be followed for approximately 1000 miles, realizing that as mileage accumulates, gradually higher engine speeds and loads can be applied. After 2000 miles have been covered, the engine can be considered satisfactorily broken in and its full performance potential can be utilized.

4　During the break-in period, keep a very close eye on the engine oil and coolant levels. Change the engine oil and filter at 500 miles and again at 2000 miles to ensure that the minute metal particles normally generated during break-in are removed.

Chapter 3 Cooling system

Contents

Specifications

General

Coolant type/capacity	See Chapter 1
Radiator cap valve opening pressure	13 psi
Cooling system test pressure	14 psi

Thermostat

Begins to open (approx.)...........................	180 °F (82 °C)
Fully open (approx.)	203 °F (95 °C)
Opening width	0.310 in (8 mm)

Torque specifications

	Ft-lb	Nm
Thermostatic switch	11	15
Coolant temperature sending unit	11	15

1 General information

The Yamaha Vision is equipped with a liquid cooling system which utilizes a water/antifreeze mixture to carry away excess heat produced during the combustion process. The cylinders are surrounded by water jackets, through which the coolant is circulated by the water pump. The pump is mounted in the right-hand crankcase cover and is driven by a gear and shaft from a crankshaft mounted gear. The hot coolant passes up through flexible hoses to the top of the radiator, which is mounted on the frame downtubes to take advantage of maximum air flow. The coolant then passes down through the radiator core, where it is cooled by the passing air, through the thermostat, then to the water pump and engine where the cycle is repeated.

An electric fan, mounted behind the radiator and automatically controlled by a thermostatic switch, provides a flow of cooling air through the radiator when the motorcycle is not moving. Under certain conditions, the fan may come on even after the engine is stopped, and the ignition switch is off, and may run for several minutes.

The entire system is sealed and pressurized. The pressure is controlled by a valve which is part of the radiator cap. By pressurizing the coolant, the boiling point is raised, which prevents premature boiling of the coolant. An overflow hose, connected between the radiator and reservoir tank, directs coolant to the tank when the radiator cap valve is opened by excessive pressure. The coolant is automatically siphoned back to the radiator as the engine cools.

Many cooling system inspection and service procedures are considered part of routine maintenance and are included in Chapter 1. **Caution:** *Do not remove the radiator cap, especially when the engine and radiator are hot. Scalding hot coolant and steam may be blown out under pressure, which could cause serious injury. To open the radiator cap, remove the plastic radiator cover (it is held in place with four screws). When the engine has cooled, place a thick rag, like a towel, over the radiator cap; slowly rotate the cap counterclockwise to the first stop. This procedure allows any residual pressure to escape. When the steam has stopped escaping, press down on the cap while turning counterclockwise and remove it.*

2 Coolant level — check

Refer to Chapter 1 for this procedure.

3   Cooling system — check

This procedure is considered part of routine maintenance and is included in Chapter 1.

4   Radiator cap — check

If problems such as overheating and loss of coolant occur, check the entire system as described in Chapter 1. The radiator cap opening pressure should be checked by a dealer service department or service station equipped with the special tester required to do the job. If the cap is defective, replace it with a new one.

5   Electric fan — check

Refer to Chapter 8, *Electrical system*, for this procedure.

6   Electric fan thermostatic switch — removal, check and installation

Note: *The engine must be completely cool before beginning this procedure.*

Removal
1   The thermostatic switch is threaded into the large aluminum T-fitting in the upper radiator hose (photo). It must be removed to be checked properly. Before removing the switch, check the electric fan and relay (see Chapter 8) and all wiring and connections.
2   Refer to Chapter 1 and drain the coolant from the radiator. It is not necessary to drain the cylinder water jackets.
3   Carefully separate the wire from the switch, then remove the switch from the T-fitting. Place a rag or paper towels under the switch to catch any coolant that may run out of the mounting hole. **Note:** *Handle the switch carefully. If it is dropped or receives a strong blow, it must be replaced with a new one.*

Check
4   Refer to Chapter 8 for the thermostatic switch checking procedure. If it is defective, replace it with a new one.

Installation
5   Carefully thread the switch into the T-fitting and tighten it to the specified torque.
6   Attach the wire to the switch terminal, then refer to Chapter 1 and refill the cooling system. Start the engine and let it warm up, then check for leaks.
7   If a new thermostatic switch was installed, make sure the fan operates normally.

7   Coolant temperature sending unit — removal, check and installation

Note: *The engine must be completely cool before beginning this procedure.*

Removal
1   The sending unit is threaded into the large aluminum T-fitting in the upper radiator hose (photos). It must be removed to be checked properly. Before removing the unit, check the coolant temperature gauge (see Chapter 8) and all wiring and connections.
2   Refer to Chapter 1 and drain the coolant from the radiator. It is not necessary to drain the cylinder water jackets.
3   Carefully separate the wire from the sending unit, then remove the unit from the T-fitting. Place a rag or paper towels under the T-fitting to catch any coolant that may run out of the mounting hole. **Note:** *Handle the sending unit carefully. If it is dropped or receives a strong blow, it must be replaced with a new one.*

Check
4   Refer to Chapter 8 for the sending unit checking procedure. If it is defective, replace it with a new one.

6.1   The thermostatic switch (1) and coolant temperature sending unit (2) are threaded into the cooling system T-fitting under the fuel tank

Installation
5   Carefully thread the unit into the T-fitting and tighten it to the specified torque.
6   Attach the wire to the sending unit terminal, then refer to Chapter 1 and refill the cooling system. Start the engine and let it warm up, then check for leaks.
7   If a new sending unit was installed, make sure the temperature gauge operates normally.

8   Coolant temperature gauge — check

Refer to Chapter 8, *Electrical system,* for the coolant temperature gauge checking procedure.

9   Thermostat — removal, check and installation

Note: *The engine must be completely cool before beginning this procedure.*

Removal
1   If the thermostat is functioning properly, the coolant temperature gauge should rise to the normal operating temperature quickly and then stay there, only rising above the normal position occasionally when the engine gets unusually hot. If the engine does not reach normal operating temperature quickly, or if it overheats, the thermostat should be removed and checked or replaced with a new one.
2   Refer to Chapter 1 and drain the entire cooling system. When the thermostat cover is removed to drain the radiator, the thermostat will be visible in the water pump housing.
3   Withdraw the thermostat and rubber gasket from the water pump.

Check
4   Remove any coolant deposits, then visually check the thermostat for corrosion, cracks and other damage. If it was open when it was removed, it is defective. Check the rubber gasket for cracks and other damage.
5   To check the thermostat operation, submerge it in a container of water along with a thermometer. The thermostat should be suspended so it does not touch the container.
6   Gradually heat the water in the container with a hotplate or stove and check the temperature when the thermostat first starts to open.
7   Continue heating the water and check the temperature when the thermostat is fully open.
8   Lift the fully open thermostat out of the water and measure the distance the valve has opened.
9   Compare the opening temperature, the fully open temperature and

the valve travel to the Specifications.
10 If these Specifications are not met, or if the thermostat does not open while the water is heated, replace it with a new one.

Installation

11 Slide the thermostat and rubber gasket into the water pump housing with the spring facing in (the water pump has two grooves which the sides of the thermostat must fit into). Refer to Chapter 1 and make sure the breather and rubber gasket are properly positioned, then refill the cooling system as described there.

10 Coolant reservoir — removal and installation

1 If leaks or other damage which require removal of the coolant reservoir should occur, simply separate the top hose from the reservoir, remove the mounting screw, lift straight up and tilt it back, then detach the lower hose. Be careful not to spill coolant on the painted surfaces of the frame.
2 Installation is the reverse of removal. Make sure the hoses are securely attached (hose clamps in place) and add coolant to the reservoir before running the engine.

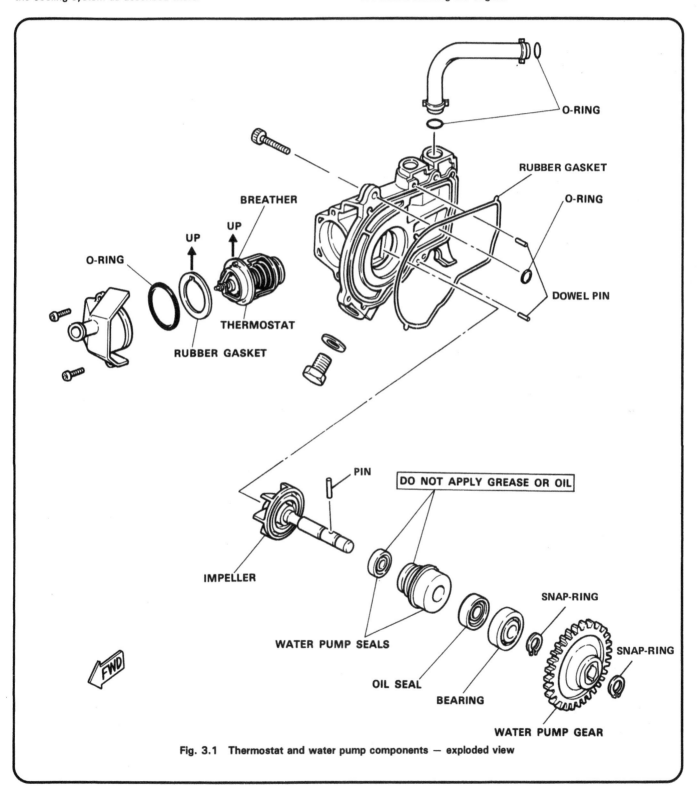

Fig. 3.1 Thermostat and water pump components — exploded view

11   Cooling system — draining, flushing and refilling

Refer to Chapter 1 for this procedure.

12   Radiator — removal and installation

Note: *The engine must be completely cool before beginning this procedure.*

1   Refer to Chapter 1 and drain the entire cooling system. If the coolant is in good condition, or relatively new, it can be reused.

2   Refer to Chapter 4 and remove the fuel tank from the motorcycle. Unplug the cooling fan wiring connector.

3   Loosen the hose clamp and separate the lower radiator hose from the radiator (photo). If it is difficult to remove, grasp the hose with water pump pliers and twist it around the radiator fitting until it can be pulled free.

4   Remove the top and right side radiator mounting bolts (photo); leave the left side bolt in place.

5   Carefully pull the right side of the radiator forward until the upper radiator hose, the bypass hose and the overflow hose are exposed, then separate them from the radiator. Expand the hose clamps with pliers (or loosen the screw) and slide them down the hoses, then pull the hoses off. If they are difficult to remove, grasp them with a pair of water pump pliers and twist them around the fittings until they can be pulled free.

6   Support the radiator, remove the remaining mount bolt and separate the radiator and cooling fan from the frame.

7   If the radiator is to be repaired, pressure checked or replaced, detach the electric fan by removing the three mounting bolts.

8   Carefully examine the radiator for evidence of leaks and damage. It is recommended that any necessary repairs be performed by a reputable radiator repair shop.

9   If the radiator is clogged, or if large amounts of rust or scale have formed, the repair shop will also do a thorough cleaning job.

10   Make sure the spaces between the cooling tubes and fins are clear. If necessary, use compressed air or running water to remove anything that may be clogging them. If the fins are bent or flattened, straighten them very carefully with a small screwdriver.

11   Installation is basically the reverse of removal. If the rubber hoses are damaged or deteriorated, now would be a logical time to replace them with new ones. Also, replace the spring type hose clamps with stainless steel, worm drive screw-type clamps.

12   After the installation is complete, refill the cooling system (see Chapter 1), start the engine and check carefully for leaks.

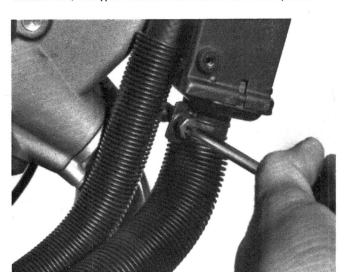

12.3   Loosening the lower radiator hose clamp

12.4   Removing the right side radiator mounting bolt

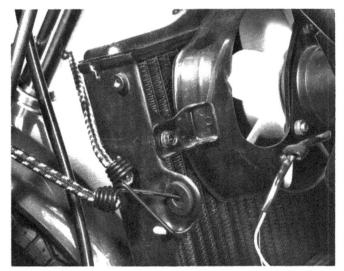

13.4   Hold the radiator forward, as shown, and remove the cooling fan mounting bolts

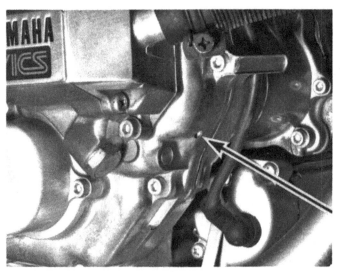

14.1   If coolant runs out the small hole in the crankcase cover (arrow) the water pump seals are leaking

13   Electric fan — removal and installation

Note: *The engine must be completely cool before beginning this procedure.*

1   Remove the fuel tank (refer to Chapter 4 if necessary) and unplug the electric fan wiring connector.

2   Remove the four mounting screws and separate the plastic cover from the radiator.

3   Remove the top and left side radiator mounting bolts; leave the right side bolt in place.

4   Carefully push the left side of the radiator forward and hold it there with an elastic cord or a length of rope or wire (photo).

5   Remove the three fan mounting bolts and separate the fan from the radiator. Be careful not to dent or bend the radiator cooling fins.

6   To separate the motor from the bracket, hold the fan and remove the nut, then slip the fan and washers off the motor shaft. Remove the three mounting screws and detach the motor.

7   Installation is the reverse of removal.

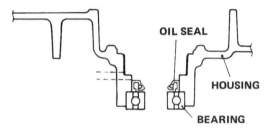

Fig. 3.2   The oil seal and bearing can be driven out after the crankcase cover water pump seal has been removed (Sec 14)

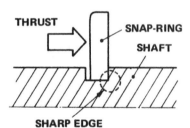

Fig. 3.3   The snap-rings on the water pump impeller shaft should be installed with the sharp edge facing away from the impeller (Sec 14)

14.4   Thermostat housing mounting screws (arrows)

14   Water pump — removal, overhaul and installation

Removal

1   **Note:** *Do not remove and disassemble the water pump unless the coolant overheats and/or runs out the small hole in the right crankcase cover (photo). The engine must be completely cool before beginning this procedure.* Refer to Chapter 1 and drain the engine oil (Section 24) and the cooling system (Section 23). It is not necessary to remove the oil filter.

2   Slide the rubber gasket and the thermostat out of the housing (if you have not already done so).

3   Separate the bypass hose from the right engine crankcase cover. To do this, expand the hose clamp with pliers and slide it up the hose. Grasp the hose (at the fitting) with a water pump pliers, twist it around the fitting and pull it free (photo).

4   Remove the thermostat housing mounting screws (photo) and separate the housing and metal coolant pipes from the engine. Be careful not to lose the two small dowel pins or the O-ring mounted in the crankcase cover.

5   To provide enough clearance to separate the right crankcase cover from the engine, the rear brake pedal must be removed. Turn the rear brake pedal free play adjusting nut counterclockwise until as much slack

14.3   Separating the bypass hose from the right crankcase cover

14.8   Remove the outer snap-ring (arrow), then separate the gear from the impeller shaft

14.9a   Removing the pin from the impeller shaft

14.9b   Remove the inner snap-ring (arrow) and carefully slide the impeller and shaft out of the cover

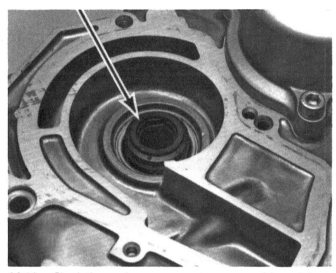

14.11a   Check the crankcase cover water pump seal (arrow) for wear and damage

14.11b   Check the impeller seal (arrow) for wear and cracks

as possible is present in the pedal and brake rod. Remove the pedal clamp bolt and pull the pedal off the shaft.

6   Place newspapers, cardboard or rags under the engine, then remove the right crankcase cover screws. Two different length screws are used so keep track of their locations.

7   Separate the crankcase cover from the engine (it may be necessary to tap it gently with a soft-faced hammer to break the gasket seal). Do not lose the dowel pins that align the cover with the engine.

Overhaul

8   Remove the outer snap-ring from the water pump impeller shaft, then lift off the gear (photo).

9   Remove the pin from the shaft (photo), then remove the inner snap-ring (photo).

10   Carefully slide the impeller and shaft out of the cover, then remove any deposits from the impeller and the pump housing. Check the impeller and shaft for cracks, wear and other damage.

11   Check the water pump seals (photos) for wear, cracks, distortion and other damage. If coolant has been discharged from the small hole in the right crankcase cover, the water pump seals have been leaking and should be replaced with new ones. If they require replacement, take the crankcase cover and impeller to a Yamaha dealer service department to have the old seals removed and new ones installed. A special seal installation tool and hydraulic press are required to properly in-

stall the seal in the crankcase cover so don't attempt to do it yourself. The impeller seal can be removed and installed without any special tools, but the seals must be replaced as a set so let the dealer service department handle it.

12   Check the oil seal for cracks and excessive wear (the seal lip should be relatively sharp and pliable). Check the bearing (photo) for smooth operation and excessive play. If the bearing and/or oil seal are damaged or worn, they must be replaced with new parts.

13   The bearing and seal can be removed and installed without any special tools, but the crankcase cover water pump seal must be removed in order to drive out the bearing and oil seal. As a result, it may be a good idea to leave the entire operation to a dealer service department.

14   If you decide to replace the bearing and oil seal yourself, drive out the water pump seal from the bearing side. Turn the crankcase cover over and drive out the oil seal and bearing. Use a deep socket or section of pipe (with an outside diameter the same as the oil seal's) and a hammer.

15   Install the new bearing, by tapping on the *outer* race only, until it bottoms against the shoulder in the crankcase cover hole. Use a deep socket with an outside diameter the same as the outer race's. **Note:** *Do not hammer on or apply pressure to the bearing inner race, as damage to the bearing will result.*

16   Turn the cover over and install the new oil seal with the manufacturer's marks or numbers facing up. **Note:** *Be careful not to damage*

14.12 Check the water pump impeller shaft bearing (arrow) for smooth operation and excessive play (the oil seal is located directly behind the bearing)

14.17 Lubricate the impeller shaft before installation but do not get grease on the sealing surface

14.18a The dowel pins and O-ring (arrows) must be in place before installing the thermostat housing

14.18b The brake pedal and shaft have index marks (arrows) that should be aligned as the pedal is installed

the seal lip and make sure the spring does not slip out of the back of the seal.
17 Reassemble the water pump by reversing the disassembly procedure. Apply a very light coat of grease to the impeller shaft (photo) before sliding it into place. **Note:** *Be very careful not to get grease on the water pump seals.* Check the snap-rings for damage and distortion before installing them and make sure they are installed with the sharp edge facing out (see accompanying illustration).

Installation
18 Installation of the water pump is basically the reverse of removal but note the following:
 a) The gasket sealing surfaces of the crankcase and cover must be clean before reattaching the cover. Use a gasket scraper to

carefully remove all traces of the old gasket; always use a new gasket and make sure the dowel pins are in place before installing the cover. Tighten the cover screws evenly and securely following a criss-cross pattern.
 b) Check the O-rings on the metal coolant pipes and replace them with new ones if necessary.
 c) Make sure the dowel pins and O-ring are in place (photo) and check the rubber gasket before installing the thermostat housing. The metal coolant pipes must be in place in the housing before it is attached to the engine.
 d) When reinstalling the brake pedal, be sure to line up the index marks on the pedal and shaft (photo). Tighten the bolt securely and adjust the brake by referring to Chapter 1.

Chapter 4 Fuel system

Contents

Specifications

General

	RJ	RK
Carburetor type	Mikuni BD34	Mikuni BD36

Jet sizes

Main jet		
Front carburetor	No. 122.5	No. 130
Rear carburetor	No. 127.5	No. 130
Main air jet	1.8	1.8
Throttle valve	No. 120	No. 130
Pilot jet	No. 60	No. 50
Pilot air jet	No. 130	No. 140
Inlet needle valve seat	1.8	2.0

Carburetor adjustments

Pilot screw (idle fuel/air mixture)	
US models	Factory preset
UK models	2-1/2 turns out
Float level	0.79 ± 0.040 in (20 ± 1 mm)
Float height	1.72 ± 0.040 in (36 ± 1 mm)

1 General information

The fuel system consists of the fuel tank, the fuel pump and pressure regulator, the carburetors and the connecting lines, hoses and controls.

The carburetors on this motorcycle are automotive-type downdraft Mikunis with butterfly-type throttle valves. An accelerator pump is used to eliminate hesitation when the throttle is opened suddenly at low speeds. RK models also have an accelerator control valve attached to the accelerator pump and an air filter assembly attached to the front float bowl vent tube (see Fig. 4.1).

The fuel pump, which is operated by vacuum pulses in the intake manifold, delivers gas from the tank to the carburetors and the regulator directs excess fuel through a special circuit back to the fuel pump inlet.

Many of the fuel system service procedures are considered routine maintenance items and for that reason are included in Chapter 1.

2 Air filter – servicing

Refer to Chapter 1, *Tune-up and routine maintenance*, for this procedure.

3 Fuel system – check

A general fuel system checking procedure is included in Chapter 1, *Tune-up and routine maintenance*.

4 Fuel petcock/filter – servicing

The fuel petcock and filter servicing procedures are covered in detail in Chapter 1.

5 Throttle operation/grip free play – check and adjustment

Refer to Chapter 1, *Tune-up and routine maintenance*, for this procedure.

6 Fuel tank – removal and installation

1   The fuel tank is held in place at the forward end by two cups, one on each side of the tank, which slide over two rubber buffers on the frame. The rear of the tank rests on a bracket and rubber saddle placed across the frame top tubes and is attached by a plate, rubber cushion and bolt which fit through a flange projecting from the tank.
2   To remove the tank, turn the fuel petcock to the On or Reserve position, slide back the hose clamps and pull the fuel and vacuum lines (photo) off of the petcock. Raise the seat and remove the bolt, steel plate and rubber cushion. After removing the plate and the bolt, the rear of the tank should be lifted up and pulled back until the cups clear the rubber buffers, then the tank can be lifted carefully away from the machine.

3   When replacing the tank, reverse the above procedure. Make sure the tank seats properly and does not pinch any control cables or wires. If difficulty is encountered when trying to slide the tank cups onto the buffers, a small amount of light oil should be used to lubricate them.

7   Idle speed — adjustment

Refer to Chapter 1, *Tune-up and routine maintenance*, for the procedure to follow when adjusting the idle speed.

8   Idle fuel/air mixture adjustment — general information

1   Due to the increased emphasis on controlling motorcycle exhaust emissions, certain governmental regulations have been formulated which directly affect the carburetion of this machine. In order to comply with the regulations, carburetors which are not adjustable (in terms of the idle fuel/air mixture) have been installed. These carburetors are designed to supply leaner fuel/air mixtures, especially at idle, which reduce exhaust emissions.
2   The pilot screw, which controls the fuel/air mixture at idle, is preset at the factory and sealed off with a metal plug (so it cannot be tampered with).
3   If the engine runs extremely rough at idle or continually stalls, and if a carburetor overhaul does not cure the problem, take the motorcycle to a Yamaha dealer service department. The dealer can replace and adjust the pilot screw, if necessary, to restore idle and low speed performance.

9   Carburetor overhaul — general information

1   Poor engine performance, hesitation, hard starting, stalling, flooding and backfiring are all signs that major carburetor maintenance may be required.
2   Keep in mind that many so-called carburetor problems are really not carburetor problems at all, but mechanical problems within the engine or ignition system malfunctions. Try to establish for certain that the carburetors are in need of maintenance before beginning a major overhaul.
3   Check the fuel filter, the fuel lines, the gas tank cap vent, the intake manifold hose clamps, the vacuum hoses, the air filter element, the cylinder compression, the spark plugs and ignition timing, the carburetor synchronization and the fuel pump before assuming that a carburetor overhaul is required.
4   Most carburetor problems are caused by dirt particles, varnish and other deposits which build up in and block the fuel and air passages.

Also, in time, gaskets and O-rings shrink and cause fuel and air leaks which lead to poor performance.
5   When the carburetor is overhauled, it is generally disassembled completely and the parts are cleaned thoroughly with solvent and dried with compressed air. The fuel and air passages are also blown through with compressed air to force out any dirt that may have been loosened but not removed by the solvent. Once the cleaning process is complete, the carburetor is reassembled using new gaskets, O-rings and, generally, a new inlet needle valve and seat.
6   Before dismantling the carburetors, make sure you have a carburetor rebuild kit (which will include all necessary O-rings and other parts), some new solvent, a supply of rags, some means of blowing out the carburetor passages and a clean place to work. It is recommended that only one carburetor be overhauled at a time to avoid mixing up parts.
7   **Caution:** *Gasoline is extremely flammable and extra precautions must be taken when working on any part of the fuel system. Do not smoke or allow open flames or bare light bulbs near the work area. Also, do not work in a garage if a natural gas type appliance with a pilot light is present.*

10   Carburetors — removal and installation

1   Refer to the appropriate Section in this Chapter and remove the fuel tank from the motorcycle.
2   Remove the rubber strap from the airbox, then separate the front crankcase vent hose from the cylinder head cover fitting. Separate the rear vent hose from the airbox by sliding back the hose clamp and pulling on the hose.
3   Loosen the clamp band screws that attach the airbox to the carburetors, then lift off the airbox.
4   Loosen the clamp screw (photo) and separate the choke cable and housing from the carburetors.
5   Detach the overflow/vent tubes from the left side of the carburetors (photo). Remove the YICS chamber mounting bolt.
6   Pull the cable housing out of the carburetor bracket, wedge the throttle in the wide open position (photo), then disconnect the throttle cable from the linkage.
7   Slide back the hose clamp and pull the fuel pump vacuum hose off of the intake manifold fitting, then loosen the carburetor-to-intake manifold clamp band screws (photo). Lift up on the carburetors to disengage them from the intake manifolds, then work them out the left side of the frame (front end first).
8   Installation is basically the reverse of removal, but be sure to slip the carburetors through the frame rear end first or problems will develop. Also, be sure to seat the carburetors squarely in the intake manifolds and tighten the clamp band screws securely.
9   Refer to Chapter 1 and adjust the throttle grip/cable free play. To adjust the choke cable, move the housing in the clamp until there is

6.2   Disconnect the fuel and vacuum lines (arrows) from the petcock before attempting to remove the tank

10.4   Loosen the clamp screw (arrow) to remove the choke cable

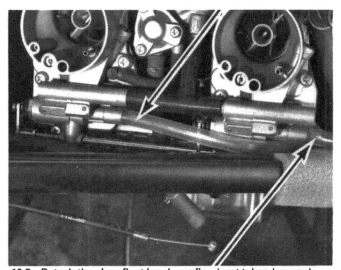

10.5   Detach the clear float bowl overflow/vent tubes (arrows) from the left side of the carburetors (RK models have an air filter assembly attached to the front carburetor with a screw and bracket)

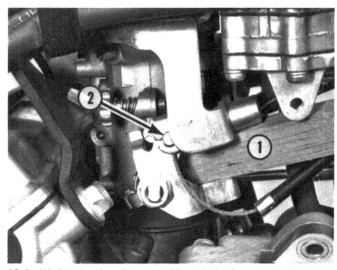

10.6   Wedge the throttle open with a block of wood (1), then slip the cable end (2) out of the lever by passing the cable through the slot

10.7   Disconnect the vacuum hose (1) from the intake manifold fitting, then loosen the clamp band screws (2) and remove the carburetors

11.2   Loosen the screws (1), then slide the starter link (2) out of the carburetor mounts

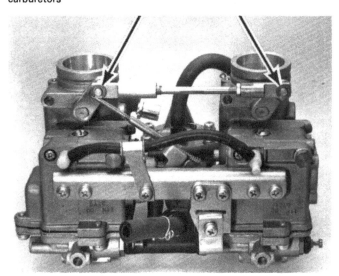

11.3   The synchronizing rod circlips (arrows) can be removed with a small screwdriver (do not distort them)

11.4   Removing the upper bracket mounting screws (4 total)

free play in the inner cable when the choke plungers in the carburetors are closed. Make sure the plungers open when the choke lever is operated.

11   Carburetors – disassembly, cleaning and inspection

Note: *Always disassemble one carburetor at a time to avoid mixing up the parts. The carburetor shown in the accompanying photos is from an RJ model.*

1   Before disassembling the carburetor, thoroughly clean the exterior with solvent to remove any accumulated dirt and grime, then let it dry. As the carburetor is disassembled, lay the parts out in the order removed on a clean shop towel or newspapers.

Disassembly

2   Loosen the two starter link screws and separate the starter link and levers from the carburetors (photo).
3   Remove the synchronizing rod circlips and separate the rod from the throttle levers (photo). **Note:** *Do not loosen the locknut on the rod*

and do not change the position of the rod in relation to the ends.

4   Remove the upper bracket mounting screws and separate the bracket and fuel pump from the carburetors (photo). The T-shaped fuel line and the pressure regulator vent hose (RJ only) can be carefully pulled off of their fittings.
5   Remove the lower bracket mounting screws (photo) and separate it from the carburetors. Carefully pull the accelerator pump fuel line and the float bowl drain line off of the carburetor fittings. Try not to crack or distort them in the process.
6   Loosen the upper body mounting screws gradually, following a crisscross pattern, then remove them (photo). Carefully separate the upper body from the main body. Do not pry between them, as damage to the gasket sealing surfaces will result.
7   Carefully tap out the float pivot pin with a pin punch and separate the float and inlet needle valve from the upper body (photo).
8   Remove the clamp screw and separate the clamp and inlet needle valve seat from the upper body (photo).
9   Using a 14 millimeter box end wrench, loosen the starter plunger fitting (photo). Remove the fitting and separate the starter plunger and spring from the upper body.
10   If working on the rear carburetor, carefully pry the accelerator pump rod out of the throttle lever with a small screwdriver (photo).

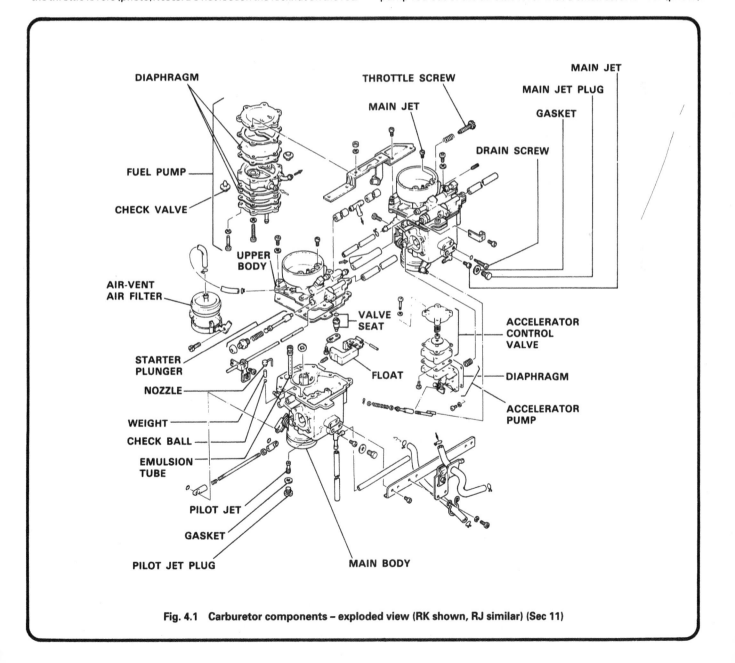

Fig. 4.1   Carburetor components – exploded view (RK shown, RJ similar) (Sec 11)

Pivot the rod and accelerator pump lever back and remove the four pump cover screws. Lift off the cover and remove the diaphragm and spring. **Note:** *The accelerator pump rod is factory preset and locked in place with Locktite. Do not attempt to disassemble or adjust it.* On RK models, disassemble the accelerator control valve (see Fig. 4.1).

11   Refer to the accompanying illustrations and remove the pilot air jet from the upper body, the float bowl drain screw, main jet plug and main jet from the side of the main body and the pilot jet plug and pilot jet from the bottom of the main body (photos). Do not remove the accelerator pump nozzle and related components unless the nozzle is obviously restricted. It is press fit into the main body and can be removed with a pliers.

12   Remove the clamp screw and separate the secondary venturi and washer from the main body (photo).

Cleaning

13   Submerge the metal components in a petroleum based solvent for approximately thirty (30) minutes. **Note:** *Do not soak the parts in caustic carburetor cleaners.*

14   After the carburetor has soaked long enough for the solvent to loosen and dissolve most of the varnish and other deposits, use a brush to remove the stubborn deposits. Rinse it again with clean solvent, then dry it with compressed air. Blow out all of the fuel and air passages in the main and upper body. *Never clean the jets or passages with a piece of wire or a drill bit, as they will be enlarged, causing the fuel and air metering rates to be upset.*

Inspection

15   Check the carburetor body for cracks and other damage. Make sure the throttle shaft turns freely without excessive play. if it is worn and sloppy, a new carburetor is in order.

Make sure the bypass holes and the pilot outlet in the carburetor bore are clear (photo).

16   Examine the floats for distortion and make sure the metal pivot section is firmly bonded to the plastic floats. Check the inlet needle valve and seat for nicks and a pronounced groove or ridge on the sealing surfaces (photo). If there is evidence of wear, the needle and seat should be replaced with new parts (always replace the needle and seat as a set). Check the float pivot pin and its bores for wear (photo). If the pin is a sloppy fit in the bores, excessive amounts of fuel will enter the float bowl and flooding will occur. Check the O-ring (photo) on the inlet needle valve seat. If it is cracked or distorted, replace it with a new one.

17   If working on the rear carburetor, check the accelerator pump/control valve spring(s) and diaphragm(s) (photo). If a diaphragm is torn or a spring is distorted, replace them as a set.

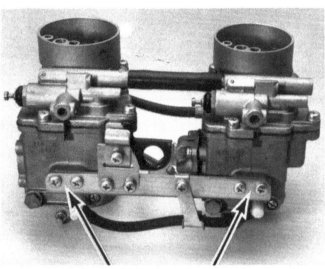

11.5   Lower bracket mounting screws (4 total)

11.6   Carburetor upper body mounting screws (6 total)

11.7   Removing the float pivot pin (mount the upper body in a vise, but be very careful not to crack or distort it)

11.8   Remove the screw (arrow) and separate the clamp and inlet needle valve seat from the upper body

11.9   The starter plunger fitting (arrow) is brass so use a box end wrench to loosen it

11.10   Pry the accelerator pump rod (arrow) out of the throttle lever with a screwdriver

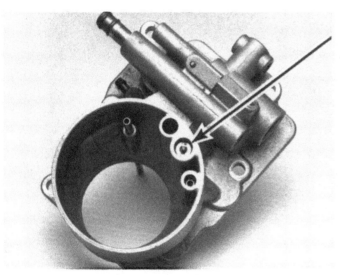

11.11a   Remove the pilot air jet (arrow), . . .

11.11b   . . . the float bowl drain screw (1), the main jet plug and main jet (2). . .

11.11c   . . . and the pilot jet plug and pilot jet (arrow) from the main body before cleaning it with solvent

11.12   The secondary venturi is wedged into the carburetor bore by the clamp screw (arrow)

11.15   Check the small pilot outlet and bypass holes (arrow) to make sure they are clear

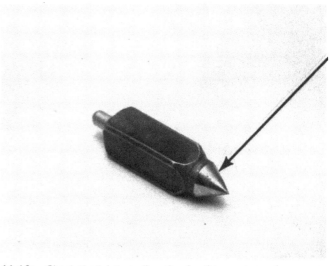

11.16a   Check the inlet needle valve for the presence of a groove in the area indicated by the arrow

11.16b   Check the float pivot pin bores (arrows) for wear

11.16c   The O-ring on the inlet needle seat (arrow) must be in good condition to prevent excess fuel from entering the float bowl

11.17   Check the accelerator pump diaphragm (arrow) very carefully for pinholes and minute tears

11.18   The starter plunger seal (arrow) must be in good condition or fuel may leak past it and cause an overly rich mixture

18   Check the end of the starter plunger. Make sure the rubber seal (photo) is in good condition. If it is cracked, distorted or deeply grooved, replace the plunger with a new one.

19   Check the secondary venturi gasket for damage (photo). If it is damaged or severely compressed, replace it with a new one.

20   Refer to Section 12 for the carburetor reassembly procedure. Repeat the disassembly, cleaning and inspection process for the remaining carburetor after the first carburetor is reassembled.

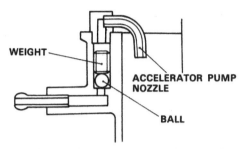

Fig. 4.2   The accelerator pump nozzle is press fit into the main body (if it is removed, make sure the check ball and weight are installed as shown) (Sec 12)

12   Carburetors — reassembly

Note: *When reassembling the carburetors, be sure to use the new O-rings, gaskets and other parts supplied in the rebuild kit.*

1   Carefully thread the pilot jet into the hole in the bottom of the main body, then install the pilot jet plug and gasket.

2   Thread the main jet into the side of the main body, then install the main jet plug and gasket. **Note:** *The main jet size is stamped into the face (photo). Do not interchange the jets between the carburetors and do not confuse the main jet with the pilot air jet (refer to the Specifications for the size).*

3   Lubricate the O-ring with multi-purpose grease, then install the float bowl drain screw in the main body.

4   Install the pilot air jet into the hole in the upper body.

5   Attach the rubber gasket to the secondary venturi, then install the venturi into the carburetor bore in the main body. The small tab on the side of the venturi must face up (photo). Use a thread locking compound on the clamp screw and tighten it securely.

6   If you are working on the rear carburetor, carefully install the accelerator pump spring and diaphragm, then position the pump cover and install and tighten the four screws. Do not overtighten them or damage to the diaphragm will result. Pivot the rod and pump lever toward the bottom of the carburetor and attach the rod to the throttle lever.

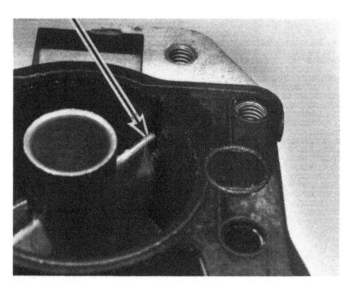

11.19   Secondary venturi gasket location

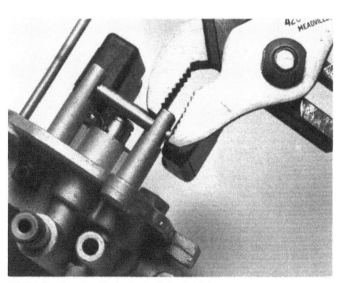

12.2   The main jet size is stamped into the face (arrow)

12.5   Install the secondary venturi with the small tab (arrow) up

12.8   Seating the float pivot pin in the bore

7   Insert the starter plunger and spring into the bore in the upper body, then install and tighten the fitting using a 14 millimeter box end wrench. Make sure the rubber boot is in position on the fitting.
8   Lubricate the O-ring on the inlet needle valve seat with multi-purpose grease, then carefully install the valve seat in the upper body. Position the clamp, then install and tighten the clamp screw. Slip the inlet needle valve into the valve seat, then hold the float in position and install the pivot pin. Using pliers, carefully squeeze the pivot pin until it is completely seated in the bore (photo).
9   Check the float height to make sure it is correct. **Note:** *Adjustment is not normally required unless the float is defective or bent.* Hold the carburetor upper body at approximately a 45-degree angle (with the float pivot pin at the top) and measure the distance from the very top of the float to the gasket surface (photo). Compare the results to the Specifications. If the float height is incorrect, separate the float from the upper body and carefully bend the tang that contacts the inlet needle valve (photo). **Note:** *Do not, under any circumstances, bend the individual float arms to make this adjustment.* Reattach the float and recheck the height.
10   At this point, if you are working on the front carburetor, lay a new gasket in place and carefully lower the upper body into position on the main body. Install the six screws and tighten them evenly, following a criss-cross pattern. Do not overtighten them, as the threaded holes in the main body could be damaged. *If you are working on the*

rear carburetor, do not attach the upper body to the main body at this time.
11   Once both carburetors have been disassembled, cleaned, inspected and reassembled, they can be rejoined. Position the carburetors in their normal relationship to each other and install the upper and lower brackets. Tighten the mounting screws evenly and securely but do not overtighten them.
12   Attach the synchronizing rod to the throttle lever posts and install the circlips with the sharp edge facing out. Make sure the circlips are seated in their grooves. **Note:** *The locknut on the synchronizing rod should be installed next to the front carburetor.*
13   Install the accelerator pump fuel line, the float bowl drain line and the T-shaped fuel inlet line.
14   At this point the upper body should still be detached from the rear carburetor. Fill the rear carburetor float bowl with new, clean gasoline, then hold the carburetors over a drain pan and operate the throttle lever(s). Observe the fuel spray pattern at the accelerator pump nozzle in the front carburetor (it may take several strokes of the throttle levers for fuel to reach the pump). The fuel pattern should be cone shaped. If it isn't, the accelerator pump nozzle must be removed and cleaned with compressed air. **Caution:** *Gasoline is extremely flammable, so be very careful when making this check.*
15   Operate the throttle lever(s) again until the accelerator pump lever stops moving. The fuel sprayed from the accelerator pump nozzle must pass between the throttle butterfly valve and the carburetor bore wall. If it hits the throttle valve or the bore wall, adjust the angle by carefully bending the nozzle with a needle-nose pliers. Repeat the spray pattern and injection angle check and adjustment at the rear carburetor nozzle.

12.9a   Checking the float height

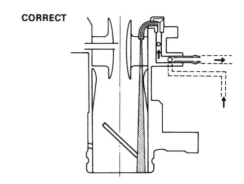

Fig. 4.3   The accelerator pump fuel spray pattern must be cone shaped and directed between the throttle butterfly valve and the carburetor bore wall (Sec 12)

12.9b   If float height adjustment is required, bend only the tang (arrow), not the float arms

12.20   Attach a section of clear fuel line to the drain fitting to check the fuel level in the float bowl

16 Install the upper body on the rear carburetor. Again, tighten the screws evenly and securely, following a criss-cross pattern, but do not overtighten them.

17 Attach the fuel pump to the upper carburetor bracket and tighten the nuts securely. Slip the pressure regulator vent hose (RJ only) onto the fitting on the front carburetor upper body.

18 Hook the starter link levers to the starter plungers. Slip the starter link into place and tighten the screws.

19 Refer to the appropriate Section in this Chapter and install the carburetors on the motorcycle, then check the fuel level as follows. **Note:** *The motorcycle must be on a level surface during this check.* Also, it may be easier to determine the fuel level if the tank is removed, placed on the seat and connected to the carburetors with a separate section of fuel line. If this is done, be sure to use a fuel line with the same inside diameter as the original and make sure the tank is held securely in place. Turn the petcock to Prime and plug the intake manifold vacuum hose before proceeding.

20 Place the motorcycle on the centerstand, then disconnect the drain line from the front carburetor. Attach a separate section of clear vinyl fuel line to the drain line fitting (photo). Hold the vinyl fuel line vertical and carefully loosen the front carburetor drain screw (do not remove it, just loosen it).

21 Turn the fuel petcock lever to the On or Reserve position (unless the tank has been removed), start the engine and let it run for approximately five minutes, then shut it off. Check the fuel level in the clear fuel line. It should be approximately 20 millimeters from the upper gasket surface of the main body (see the accompanying illustration).

22 If the fuel level is not as specified, the carburetors must be removed and the float, inlet needle valve and seat checked. Also, recheck the float height as described earlier. If no damage is found, adjust the float level slightly by carefully bending the tang on the float, then recheck the fuel level.

23 Remove the clear fuel line from the front carburetor drain fitting. Reinstall the original drain line, then remove the line from the rear carburetor drain fitting and repeat the fuel level check at the rear carburetor.

24 Refer to Chapter 1 and check/adjust the carburetor synchronization.

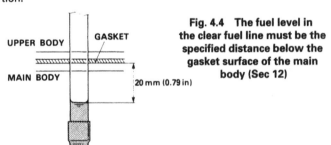

Fig. 4.4 The fuel level in the clear fuel line must be the specified distance below the gasket surface of the main body (Sec 12)

UPPER BODY   GASKET

MAIN BODY

20 mm (0.79 in)

13 Carburetors – synchronization

Since carburetor synchronization is a routine maintenance procedure, it is covered in Chapter 1.

14 Fuel pump – check

1 Refer to the appropriate Section in this Chapter and remove the fuel tank from the motorcycle. Detach the airbox by removing the strap, the crankcase vent hoses and the carburetor clamp band screws (refer to Section 10, Paragraph 2 for more details).

2 Carefully separate the T-shaped fuel line from the fuel pump outlet (leave it attached to the carburetors).

3 Attach a separate section of fuel line to the pump outlet (photo). The line should be about two (2) feet long so it can be passed between the top of the carburetors and the frame to the left side of the motorcycle. Place the free end of the hose in a plastic bottle or a can with at least a one (1) quart capacity.

4 Replace the tank (leave the airbox off) and hook up the fuel line and vacuum hose. Position the petcock lever in the Prime position.

5 Start the engine (it will run long enough on the fuel in the carburetor float bowls) and make sure a steady stream of gasoline in ejected from the fuel pump outlet. **Caution:** *Gasoline is extremely flammable, so be very careful when performing this check.*

6 If the pump output is slow or sporadic, stop the engine and check the vacuum hose between the intake manifold and the fuel pump. Look for cracks, crimps and other damage.

7 If the vacuum hose is in good condition, the pump and pressure regulator (RJ only) should be removed, disassembled and inspected.

8 When the check is complete, position the petcock lever in the On or Reserve position and remove the tank, then reinstall the airbox.

15 Fuel pump – removal and installation

1 Refer to the appropriate Section in this Chapter and remove the fuel tank from the motorcycle. Detach the airbox by removing the strap, the crankcase vent hoses and the carburetor clamp band screws (refer to Section 10, Paragraph 2 for more details). Stuff paper towels or rags into the carburetor throats in case you drop something into them.

2 Separate the YICS canister from the fuel pump by removing the single mounting bolt.

3 Separate the fuel pump vacuum hose from the intake manifold, then remove the two pump mounting nuts (photo). Push down on the pump and regulator assembly until the screws clear the carbure-

14.3 Make sure the substitute fuel line (arrow) is securely attached to the pump outlet before starting the engine

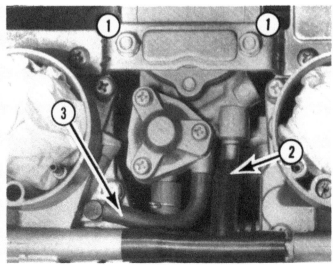

15.3 Remove the mounting nuts (1), the outlet fuel line (2) and the pressure regulator vent hose (3) (RJ only) before separating the fuel pump from the carburetors

16.1   To disassemble the fuel pump, remove the six screws; the long screws (arrows) must be reinstalled as shown (RJ model shown, RK similar)

16.2   RJ model pressure regulator cover screws (the cover is under a slight amount of spring pressure)

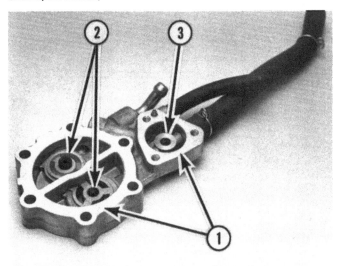

16.4   Check the gasket sealing surfaces (1), the inlet and outlet valves (2) and the pressure regulator valve seat (3) for nicks, scratches and other damage

16.6   Removing a fuel pump valve

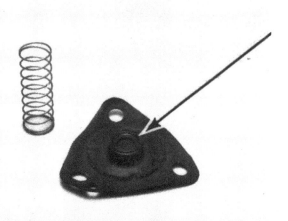

16.7   On RJ models, check the pressure regulator diaphragm valve seat (arrow) for damage that could cause leaks

16.9   On RJ models, the pressure regulator diaphragm must be installed exactly as shown, with the air nozzle (arrow) projecting through the small hole

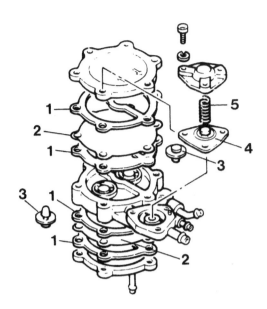

Fig. 4.5   Fuel pump components – exploded view (RJ model shown – RK models do not have a pressure regulator) (Sec 16)

1	Gasket	4	Regulator diaphragm
2	Diaphragm	5	Spring
3	Valve		

tor bracket, then separate the T-shaped outlet fuel line from the pump and the regulator vent hose (RJ only) from the carburetor fitting. Use a screwdriver positioned between the end of the hose and the fitting to pry the hoses free.

4   Installation is basically the reverse of removal. Be sure to attach the fuel, vacuum and vent hoses to the proper fittings.

16   Fuel pump – disassembly, inspection and reassembly

1   Remove the six screws (photo) and separate the covers, gaskets and diaphragms from the pump body. Do not remove the valves unless they are damaged or worn.

2   On RJ models only, remove the three screws (photo) and separate the fuel pressure regulator cover, spring and diaphragm from the pump.

3   Wash the pump body and covers in a petroleum based solvent to remove all dirt, water, rust, varnish and other deposits. Make sure the passages are clean; use a tiny brush and compressed air to remove

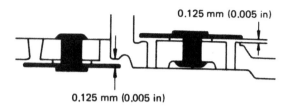

Fig. 4.6   The fuel pump valves have a designed-in gap between the plastic diaphragm and pump body seat which should not be mistaken for a defect (Sec 16)

dirt and other deposits from them.

4   Check the pump body and covers for cracks and damage to the sealing surfaces that contact the diaphragms and gaskets (photo). Check the threaded holes in the upper cover for damage. If cracks or damage are noted, replace the pump with a new one.

5   Check the gaskets and diaphragms for cracks, tears and distortion. If any damage is noted, new parts will be required.

6   Examine the valves for cracks and distortion. If in doubt as to their condition, replace them with new parts. **Note:** *As shown in the accompanying illustration, a gap of 0.005 in (0.125 mm) is provided between the valve diaphragm and the pump body seats (both valves). This is not a defect and should not be considered as such. Do not remove the valves from the pump body unless new parts are being installed.* To remove them, tuck the end of the rubber post (the end opposite the valve) into the hole with a small screwdriver (photo), then carefully pull the valve and post out from the other side. Pull only on the rubber post, not the valve. Before installing the new valves, apply petroleum jelly or very light grease to the end of the rubber post so it will slip through the hole easily.

7   Inspect the pressure regulator diaphragm and spring (if equipped) (photo). If the diaphragm is torn or the spring damaged or distorted, they should be replaced with new parts. Check the valve seat and the diaphragm seal for scratches and other damage. If the seat is damaged, the regulator will bypass fuel prematurely.

8   Assemble the diaphragms, gaskets and covers and install the screws. **Note:** *The 40 mm (longest) screws must be installed as shown in photo 16.1.* Make sure the upper and lower covers are not interchanged; the lower cover has the vacuum fitting installed. tighten the screws evenly and securely following a criss-cross pattern. Do not overtighten them or the gaskets and diaphragms will be distorted and may leak.

9   Install the regulator diaphragm (if equipped) with the air nozzle in the pump body projecting through the small hole in the diaphragm (photo). Install the spring and regulator cover, then thread the screws into the pump body and tighten them evenly and securely (do not overtighten them).

Chapter 5 Ignition system

Contents

Specifications

Ignition system fuse rating .... 10 amp

Ignition timing
Initial .... 10° BTDC at 1300 rpm
Full advance .... 38° BTDC at 4000 rpm

Ignition coil resistance
Primary winding .... 2.75 ± 10% ohms
Secondary winding .... 7.9 ± 20% K ohms

TCI pickup coil resistance .... 110 ± 10% ohms

---

1 General information

This motorcycle is equipped with a battery operated, fully transistorized, breakerless ignition system known as TCI (Transistor Controlled Ignition). The system consists of the following components:

Pickup unit and flywheel (attached to the inside of the left crankcase cover and the left end of the crankshaft)

Igniter (TCI) unit (attached to the frame under the right side cover, next to the coolant reservoir)

Battery and fuse

Coils

Spark plugs

Sidestand switch and relay

Neutral switch

Stop and main (key) switches

Primary and secondary circuit wiring

The TCI system functions on the same principle as a conventional DC ignition system with the pickup unit and igniter performing the tasks normally associated with the breaker points and mechanical advance system. As a result, adjustment and maintenance of ignition components is eliminated (with the exception of spark plug replacement).

The various switches and relays mentioned above ensure that the engine will start and run only when the sidestand is up or the sidestand is down and the transmission is in Neutral (refer to the wiring diagrams in Chapter 8).

Because of their nature, the individual ignition system components can be checked but not repaired. If ignition system troubles occur, and the faulty component can be isolated, the only cure for the problem is to replace the part with a new one. Keep in mind that most electrical parts, once purchased, cannot be returned. To avoid unnecessary expense, make very sure the faulty component has been positively identified before buying a replacement part.

---

2 Spark plugs — replacement

Refer to Chapter 1, *Tune-up and routine maintenance,* for the spark plug replacement procedure.

---

3 Ignition timing — check

Refer to Chapter 1, *Tune-up and routine maintenance,* for the procedure to follow when checking the ignition timing.

4 Sidestand switch — check

1   If the sidestand switch (located on the back side of the sidestand mounting bracket) fails, the ignition system may not operate.
2   To gain access to the switch connector, remove the right side cover, the coolant reservoir and the TCI unit. The reservoir and TCI unit can be moved aside after removing the screw that holds the top of the reservoir in place (you may have to disconnect the upper reservoir hose, but be sure to leave the lower hose in place).
3   Locate the sidestand switch wiring connector and unplug it (the connector contains two wire terminals; one is attached to a black wire and the other is attached to a blue/yellow wire).
4   Connect one ohmmeter lead to each of the wiring terminals in the connector plug attached to the sidestand switch wires. **Note:** *The easiest way to accomplish the attachment of the meter leads is to insert an automotive-type tab connector (male) (photo) into each of the connector terminals, then attach the meter leads to the protruding tab connector ends.* Place the ohmmeter selector switch in the Rx1 position.
5   With the sidestand up, the ohmmeter should read zero (0) ohms. When the sidestand is down, the ohmmeter should read infinite resistance. **Note:** *A continuity test light can be used in place of an ohmmeter. If a light is used, it should glow when the sidestand is up and go out when the sidestand is lowered.*

6   If the switch must be replaced with a new one, simply remove the two mounting screws and separate it from the sidestand bracket. Always check a new switch for proper operation after it is installed.

5 Sidestand relay — check

1   The sidestand relay is located behind the TCI unit. Refer to Section 4 and remove the TCI unit and coolant reservoir as described in Paragraph 2.
2   The relay is attached to a frame tab by a rubber cushion mount (photo). It can be identified by examining the wiring at the connector. Look for a connector with a red/white wire, a black/white wire, a blue/yellow wire and a black wire leading into it. In addition, the sidestand relay has a dab of blue paint on it and the wires are wrapped with blue tape about one inch behind the connector.
3   Pull the relay and rubber mount off the frame tab, then depress the connector lock and separate the connector from the relay.
4   Using an ohmmeter, check the resistance of the relay coil windings. Hook up the ohmmeter leads as indicated in Photo 5.4. The resistance should be 100 ohms; if it is not, replace the relay with a new one. If the resistance is as specified, proceed to Paragraph 5.
5   Connect the motorcycle's battery and an ohmmeter to the relay ter-

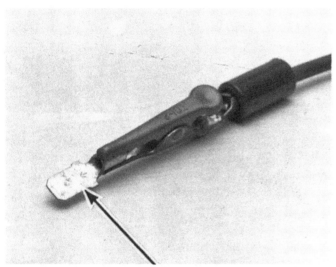

4.4a   Use an automotive-type tab connector (arrow) to tap into the wiring harness connector terminals, then attach the meter lead to the tab connector

4.4b   Checking the sidestand switch continuity with an ohmmeter (black and blue/yellow wires)

5.2   The sidestand relay (arrow) is located behind the TCI unit/coolant reservoir mount

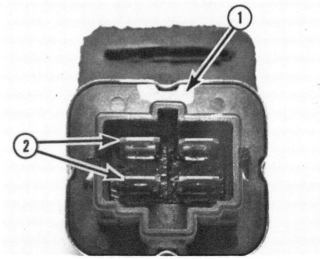

5.4   The sidestand relay is marked with blue paint (1); check the coil resistance by attaching the meter leads to the terminals labelled (2)

minals as shown in Photo 5.5 (use jumper leads with automotive tab-type female connectors to hook up the battery). With the battery *connected,* the ohmmeter should read zero (0) ohms. With the battery *disconnected*, the ohmmeter should read infinite resistance. **Note:** *A continuity test light can be used in place of an ohmmeter. If a light is used, it should glow when the battery is connected and go out when it is disconnected.*
6   If the relay fails the check, replace it with a new one.

6   Ignition system — check

1   If the ignition system is the suspected cause of poor engine performance or failure to start, a number of checks can be made to isolate the problem.
2   Make sure the ignition stop switch and the key switch are both in the Run or On position.

Engine will not start
3   Remove one of the spark plugs, hook up the plug lead and lay the plug on the engine with the threads contacting the cylinder (photo). Crank the engine over and make sure a well-defined, blue spark occurs between the spark plug electrodes.
4   If no spark occurs, substitute a new spark plug and repeat the test.

If the spark is still not satisfactory, the following checks should be made.
5   Unscrew the spark plug cap from the plug wire and check the cap resistance with an ohmmeter (photo). If the resistance is infinite, replace it with a new one.
6   Make sure all electrical connectors are clean and tight. Check all wires for shorts, opens and correct installation.
7   Check the battery voltage with a voltmeter and the specific gravity with an hydrometer (see Chapter 1). If the voltage is less than 12 volts or if the specific gravity is low, recharge the battery.
8   Check the ignition fuse and the fuse connections. If the fuse is blown, replace it with a new one; if the connections are loose or corroded, clean or repair them.
9   Refer to Section 7 and check the primary and secondary ignition coil resistance.
10   Refer to Section 9 and check the pickup coil resistance.
11   If the preceeding checks produce positive results but there is still no spark at the plug, replace the TCI unit (igniter) with a new one (Section 11). **Note:** *Double-check everything else before replacing the TCI unit. It is very easy to overlook something simple and spend a lot of money replacing components that are perfectly good.*

Engine starts but misfires
12   If the engine starts but misfires, make the following checks before deciding that the ignition system is at fault.

5.5   Checking the sidestand relay operation by connecting jumper leads (left) to a 12-volt battery

6.3   Checking for spark at the plug electrodes

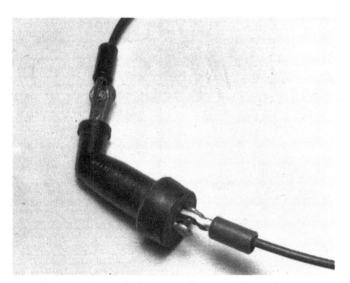

6.5   Checking the spark plug cap resistance with an ohmmeter

6.13   A simple spark gap testing fixture can be made from a block of wood, a large alligator clip, some nails, screws and wire and the cap end of an old spark plug

13   The ignition system must be able to produce a spark across a six (6) millimeter gap (minimum). A simple test fixture (photo 6.13) can be constructed to make sure that the minimum spark gap can be jumped. Make sure that the fixture electrodes are positioned six millimeters apart.

14   Hook one of the spark plug wires to the protruding test fixture electrode, then attach the fixture's alligator clip to a good engine ground (photo).

15   Crank the engine over (it may start and run on the remaining cylinder) and see if well-defined, blue sparks occur between the test fixture electrodes. If the minimum spark gap test is positive, the ignition system is functioning properly. If the spark will not jump the gap, or if it is weak (orange colored), refer to Paragraphs 5 through 11 of this Section and perform the component checks described.

16   Repeat the check for the remaining cylinder.

7   Ignition coils — check

1   In order to determine conclusively that the ignition coils are defective, they should be tested by an authorized Yamaha dealer service department which is equipped with the special electrical tester required for this check.

2   However, the coils can be checked visually (for cracks and other

damage) and the primary and secondary coil resistances can be measured with an ohmmeter. If the coils are undamaged, and if the resistances are as specified, they are probably capable of proper operation.

3   To check the coils for damage, they must be removed (see Section 8). To check the resistances, simply remove the fuel tank (Chapter 4), unplug the primary circuit wiring connector(s) and remove the spark plug wire(s) from the plug(s). The primary circuit wiring connectors are attached to the orange and red/white wires that exit the coils.

4   To check the primary coil resistance, attach one ohmmeter lead to the orange wire terminal and the other ohmmeter lead to the red/white wire terminal in the connector (photo). The easiest way to accomplish the attachment of the meter leads is to insert an automotive-type tab connector (male) into each of the connector terminals, then attach the meter leads to the protruding male connectors.

5   Place the ohmmeter selector switch in the Rx1 position and compare the measured resistance to the Specifications.

6   If the primary coil resistance is as specified, check the secondary coil resistance by disconnecting the meter lead from the red/white wire terminal and attaching it to the spark plug wire terminal (leave the ohmmeter lead attached to the orange primary wire terminal).

7   Place the ohmmeter selector switch in the Rx100 position and compare the measured resistance to the Specifications.

8   If the resistances are not as specified, the coil is probably defective and should be replaced with a new one.

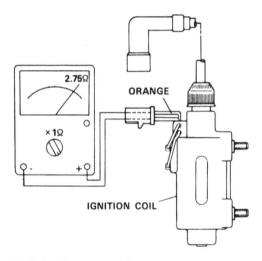

Fig. 5.1   Ohmmeter lead attachment for checking ignition *primary* coil resistance (Sec 7)

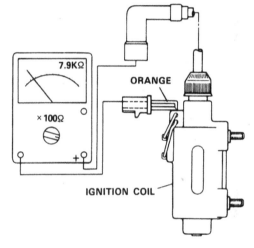

Fig. 5.2   Ohmmeter lead attachment for checking ignition *secondary* coil resistance (Sec 7)

6.14   The spark must be able to jump a 6 mm gap (minimum)

7.4   Checking the primary ignition coil resistance at the wiring connector terminals

8.3   The ignition coil mounting nuts (arrows) can be removed as the coil is supported inside the frame

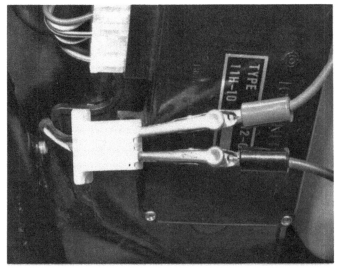

9.2   Checking the pickup coil resistance with an ohmmeter

10.4   Disconnect the clutch cable from the lever and the wires from the neutral switch and oil pressure sending unit (arrows)

10.6   Use of a simple cardboard holder will guarantee that the bolts, brackets and dowel pins are returned to their original locations

8   Ignition coils — removal and installation

1   The coils are mounted directly behind the steering head inside the frame and are attached to the gussets that span the frame tubes.
2   To remove the coils, refer to Chapter 4 and remove the fuel tank, then disconnect the spark plug wires from the plugs. Unplug the coil primary circuit wiring connectors (one for each coil).
3   Support the coil with one hand and remove the coil mounting nuts (photo), then withdraw the coil from the frame.
4   Installation is the reverse of removal. If both coils were removed, refer to the wiring diagram at the end of Chapter 8 and make sure the primary circuit wiring connectors are hooked up correctly (the connector with the gray wire attaches to the right-hand coil).

9   Pickup coils — check

1   Remove the right side cover and disconnect the pickup coil wiring harness from the TCI unit (the wiring harness consists of a red wire, a black wire and a white wire).
2   Attach one lead of an ohmmeter to the black wire terminal in the connector and the other lead to the red wire terminal (photo), then make sure the ohmmeter selector switch is in the Rx10 position and note the indicated resistance. **Note:** *The easiest way to hook up the meter leads is to insert an automotive-type tab connector (male) into each of the wire terminals in the harness connector, then attach the meter leads to the protruding tab connector ends.*
3   Move the meter lead from the red wire terminal to the white wire terminal (leave the lead attached to the black wire) and note the resistance reading on the meter.
4   In both cases, the resistance must be 99 to 121 ohms or the pickup coil assembly must be replaced with a new one (Section 10).
5   Reattach the wiring harness connector when the check is complete.

10   Pickup coils — removal and installation

1   The pickup coils are located inside the left engine crankcase cover. If they are defective, remove the coolant reservoir mounting screw, lift the reservoir and igniter out of the frame mounts, disconnect the neutral switch/oil pressure sending unit wiring harness connector (blue and black/red wires) and thread the pickup coil wiring harness out through the left side of the frame. **Note:** *You may have to remove the regulator/rectifier mounting screws so it can be moved aside to allow room for the wiring harness and connector to be withdrawn.*

10.8   Pickup coil/wire guide mounting screw locations

10.10a   Make sure the starter idler gear and shaft and the two dowel pins (arrows) are in place before installing the cover

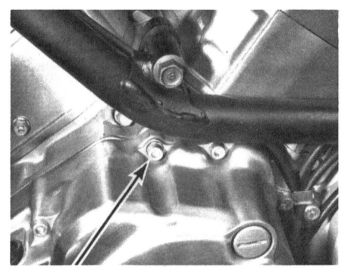

10.10b   Make sure the top cover bolt (arrow) sealing washer is installed or an oil leak will develop

11.2   Unplug the connectors (1) and remove the screw (2) to separate the TCI unit from the plastic bracket

2   Unplug the alternator stator wiring harness and thread it out to the left side of the frame also.

3   When the left engine crankcase cover is removed, oil will run out unless it is drained (see Chapter 1) or the machine is tilted to the right at about a 45° angle. *Do not lay it down flat on the right side.* Place newspapers or rags under the engine to catch any oil that may drip.

4   Pull forward on the clutch lever located on the underside of the engine and disengage the clutch cable end. Pull the cable housing out of the engine cover mount and the wire guide. Detach the wires from the neutral switch and the oil pressure sending unit on the lower left side of the engine (photo).

5   Remove the bolt and slide the shift linkage arm off the shaft.

6   Remove the crankcase cover bolts (it may be helpful to make a simple holder from a piece of cardboard (photo) to ensure that the bolts are returned to their correct locations when the cover is reinstalled).

7   Separate the cover from the engine. It may be necessary to tap it gently with a soft-faced hammer to break the gasket seal. **Caution:** *Do not use a screwdriver to pry between the cover and the engine — damage, and eventually leaks, will occur.* When the cover is loose, the alternator stator will be attracted to the magnets in the rotor, which will tend to hold the cover in place. Also, the starter idler gear shaft may stick in the cover and cause the gear to fall as the cover is removed (be prepared to catch it).

8   Note how the wiring harnesses are routed, then remove the four screws and separate the pickup coil assembly and wire guide from the crankcase cover (photo).

9   Install the new pickup coil assembly and tighten the screws securely. Use a thread locking compound on the screw threads. Be sure to route the wiring harnesses correctly and seal the grommets with Yamabond 4 or RTV-type gasket sealant. **Note:** *The grommets are shaped so that the pickup coil harness must be under the alternator stator coil wiring harness.*

10   The crankcase cover installation procedure is basically the reverse of removal. Make sure the starter idler gear and shaft (photo) are installed before the cover is replaced. Also, use a new gasket, install the two dowel pins and tighten the mounting bolts evenly and securely following a criss-cross pattern. **Note:** *The top cover bolt (photo) must have a copper sealing washer installed or oil will leak from around the bolt head.*

11   TCI unit — removal and installation

1   The TCI unit is attached to a plastic bracket behind the right side cover.

2   To remove it, unsnap the side cover, unplug the two wiring harness connectors and remove the single mounting screw (photo).

3   Installation is the reverse of removal. Do not overtighten the mounting screw — it threads into the plastic bracket. Also, make sure that the terminals are dry and corrosion-free before attaching the wiring harness connectors to the TCI unit.

Chapter 6   Frame and suspension

Contents

Specifications

Front suspension

Steering head bearings
    Number (upper and lower) 19
    Diameter .... 1/4 in (6 mm)
Front fork spring free length
    RJ .... 18.44 in (461 mm)
    RK .... 19.04 in (476 mm)
Fork oil (type, capacity and level) .... See Chapter 1

Rear suspension

Swingarm side clearance
    RJ .... 0.004 to 0.012 in (0.1 to 0.3 mm)
    RK .... 0.040 in (1.0 mm)

Torque specifications	Ft-lb	Nm
Upper fork tube pinch bolts	14	20
Lower triple clamp bolts	27	38
Shock upper mount bolt/nut	18	25
Swingarm pivot shaft (bolt)	56	78
Fork cap bolts	14	20
Fork damper rod bolt	19	26

1   General information

Yamaha's Vision utilizes a "hang-support" frame, which incorporates the engine as a structural part of the frame. This design results in lower overall weight, a lower center of gravity and easier engine removal.

The front suspension is conventional, consisting of telescopic, oil-damped, trailing axle forks. RK models have air-assisted forks.

The rear suspension is Yamaha's patented Monoshock, which is composed of a swingarm and a single hydraulically-damped shock absorber. The shock can be adjusted by changing the preload on the spring. **Note:** *The rear shock absorber cannot be rebuilt. If it wears out, it should be replaced with a new one (dispose of the old shock as described in Section 4).*

2 Frame — inspection and repair

1   The frame should not require attention unless accident damage has occurred. In most cases, frame replacement is the only satisfactory remedy for such damage. A few frame specialists have the jigs and other equipment necessary for straightening the frame to the required standard of accuracy, but even then there is no simple way of assessing to what extent the frame may have been overstressed.

2   After the machine has accumulated a lot of miles, the frame should be examined closely for signs of cracking or splitting at the welded joints. Rust corrosion can also cause weakness at these joints. Loose engine mount bolts can cause ovaling or fracturing of the mounting tabs. Minor damage can often be repaired by welding, depending on the extent and nature of the damage.

3   Remember that a frame which is out of alignment will cause handling problems. If misalignment is suspected as the result of an accident, it will be necessary to strip the machine completely so the frame can be thoroughly checked.

3 Handlebars — removal and installation

1   Look closely at how the cables and wiring harnesses are routed before removing the bars. It may be helpful to draw a simple diagram or take an instant photo to use as a reference when reinstalling the controls.

2   Pry the plastic wire holders out of the upper fork tube pinch bolts (which double as handlebar mounting bolts) and separate the wire harnesses from the clips in the bars. Remove the rear view mirrors.

3   Remove the clutch side switch assembly. It is held in place with two screws which are accessible from the front. Reassemble the switch by installing the screws finger tight and let it hang by the wire harness. Loosen the clutch lever pivot clamp bolt.

4   Remove the two bolts and separate the brake lever/reservoir assembly from the right handlebar. Carefully tie it out of the way, *but do not allow the master cylinder/reservoir assembly to hang upside down.*

5   Remove the two screws attaching the throttle to the right bar, then separate the throttle halves, disconnect the cable from the grip and slip the grip off the bar. Reassemble the throttle and let it hang by the cable.

6   Remove the upper fork tube pinch bolts (photo) and separate the handlebar assemblies from the upper triple clamp. **Note:** *The fork tube pinch bolts are very tight and may require a 1/2 inch drive, six (6) millimeter hex (Allen) head socket to remove easily.* After the left bar is removed, slide the clutch lever pivot assembly off the bar.

7   Installation is basically the reverse of removal. Completely tighten the front fork tube pinch bolts to the specified torque first, followed by the rear bolts; *the gap between the handlebar and the upper triple clamp must be at the rear (photo).*

4 Rear shock absorber — removal and installation

1   Place the motorcycle on the centerstand. Refer to Chapter 4 and remove the fuel tank, then remove the left side cover.

2   Disconnect the battery cables from the battery (negative first, followed by positive), then remove the plastic battery cover (it is held in place with two screws).

3   Raise the seat and remove the left trim piece located just below the seat. It is held in place with two bolts.

4   Push up on the starter solenoid to separate it from its rubber holder, then separate the holder from the battery case. Refer to Chapter 8 if necessary. Remove the rubber plug (opposite the lower shock mount) from the fender.

5   Refer to Chapter 5 and remove the TCI unit mounting bracket, then remove the rubber plug from the right side of the fender.

6   Remove the cotter pin (photo) from the swingarm shock mount, then support the rear wheel by slipping a block (or blocks) of wood under the tire. Drive out the pin that holds the lower part of the shock to the swingarm (it is removed by pushing from the left side to drive it out to the right).

3.6   The fork tube pinch bolts (arrows) hold the handlebars to the upper triple clamp

3.7   Tighten the front bolts first, then the rear bolts; the gap (arrow) must be at the rear

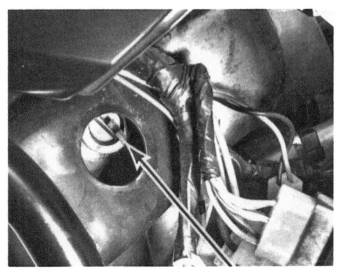

4.6   Before driving out the lower shock absorber mount pin, remove the cotter pin (arrow) from the swingarm

7   Remove the upper shock mounting nut and bolt (photo), then withdraw the shock by pulling it up and out of the frame.
8   If the shock is defective, it must be replaced with a new one. **Caution:** *Before disposing of a worn out shock absorber, release the gas pressure by carefully drilling a small (0.080 to 0.120 in) hole through the body of the shock as shown in the accompanying illustration.* **When drilling the hole, eye protection must be worn to prevent metal chips from being blown into your eyes.**
9   Installation is the reverse of removal. Be sure to tighten the upper mount bolt nut to the specified torque and install a new cotter pin in the swingarm mount. Bend the end of the cotter pin as shown in the accompanying illustration.

5   Swingarm bearings — check

1   Refer to Chapter 7 and remove the rear wheel, then remove the shock absorber (Section 4).
2   Grasp the rear of the swingarm with one hand and place your other hand at the junction of the swingarm and the frame (on the right side). Try to move the rear of the swingarm from side-to-side. Any wear (play) in the bearings should be felt as movement between the swingarm and the frame at the front. The swingarm will actually be felt to move forward and backward at the front (not from side-to-side). If any play is

4.7   Removing the upper shock absorber mounting bolt/nut

noted, the bearings should be replaced with new ones (Section 7).
3   Next, move the swingarm up and down through its full travel. It should move freely, without any binding or rough spots. If it does not move freely, refer to Section 7 for servicing procedures.

6   Swingarm — removal and installation

1   Refer to Chapter 7 and remove the rear wheel, then remove the shock absorber (Section 4).
2   Bend back the locking tab with a hammer and punch, then loosen the swingarm pivot shaft (bolt) (photo). Separate the rubber boot from the lip on the swingarm (left side). Work carefully to avoid tearing the boot.

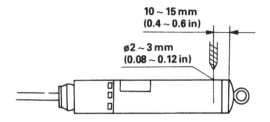

Fig. 6.1   When disposing of a worn out shock, drill a small hole in the location shown *(Wear eye protection!)* (Sec 4)

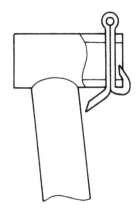

Fig. 6.2   Bend the cotter pin at the lower shock mount on the swingarm as shown to prevent it from backing out (Sec 4)

6.2   Bend back the locking tab (1), then loosen the swingarm pivot shaft (bolt) (2)

6.4   Check the swingarm bearing rollers (arrow) for wear and corrosion

3 Support the swingarm and pull the pivot shaft out, then pull straight back on the swingarm to separate it from the frame. **Note:** *Do not turn the swingarm upside down, as oil will run out of the final drive assembly vent.*

4 Remove the bearing caps, slide out the inner bushing and check the needle bearings for wear and damage (photo). Look for corrosion and make sure they roll freely. If they are worn or damaged, replace them with new ones (Section 7). Pull out the rubber bearing cap seals and remove the thrust washers.

5 Check the rubber bearing cap seals, the thrust washers, the pivot shaft and the inner bushing for wear, cracks and damage (photo). The pivot shaft is especially susceptible to corrosion. Replace any worn or damaged parts with new ones.

6 If the driveshaft or final drive components require service, the final drive assembly can be separated from the swingarm by removing the four mounting nuts and the bolt. Once the final drive assembly has been removed, the driveshaft can be withdrawn from the swingarm. Since the drivetrain components cannot be serviced by the do-it-yourselfer, it would be a good idea to take the entire swingarm assembly to a Yamaha dealer for disassembly and repair.

7 To begin installation, measure the width of the frame tube that the bearings are mounted in with a dial or Vernier caliper and record the measurement. Measure the length of the bushing in the same manner (photo) and record it as well. Using a micrometer, measure the thickness of each of the thrust washers (photo) and record the measurements. Add the thrust washer thicknesses to the frame tube width and subtract the total from the bushing length to obtain the swingarm side clearance. If it is not as specified, thicker or thinner thrust washers will have to be substituted to bring it into the specified range.

8 Once the side clearance has been checked/adjusted, liberally coat the bearings with waterproof wheel bearing grease (preferably one containing molybdenum disulfide), then slip the bushing into the bearings. Wipe off any grease that is forced out.

9 Place the thrust washers and rubber seals in the bearing caps, then

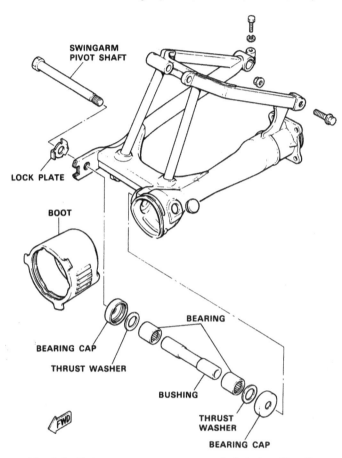

Fig. 6.3   Swingarm components — exploded view (Sec 6)

SWINGARM
PIVOT SHAFT

LOCK PLATE

BOOT

BEARING CAP

THRUST WASHER

BUSHING

BEARING

THRUST WASHER

BEARING CAP

FWD

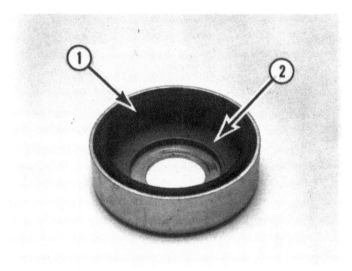

6.5   Check the rubber bearing cap seals (1) and the thrust washers (2) for wear and damage

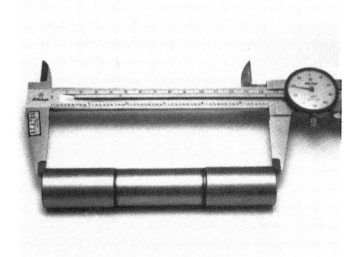

6.7a   Measuring the bushing length with a dial caliper

6.7b   Measuring the thickness of a thrust washer with a micrometer

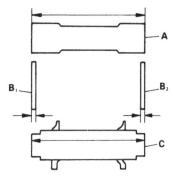

Fig. 6.4   The swingarm side clearance is equal to A (bushing length) minus $B_1 + B_2 + C$ (thrust washer thicknesses and frame tube length) (Sec 6)

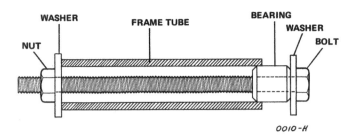

Fig. 6.5   Homemade swingarm bearing press details (Sec 7)

slip the caps over the frame tube. Make sure the thrust washers fit over the ends of the bushing.

10 Slide the swingarm into place from the rear while guiding the U-joint yoke inside the boot onto the front end of the driveshaft. You may have to turn the U-joint yoke slightly to align the splines of the yoke and shaft.

11 Support the swingarm and slide the pivot shaft and lock plate into place. The holes in the swingarm will have to be perfectly aligned with the hole in the bushing and the bearing caps. Thread the shaft into the hole in the opposite side of the swingarm.

12 Tighten the pivot shaft to the specified torque, then bend the remaining locking tab up against one of the pivot bolt head flats. Do not bend up the tab that has already been used (if both locking tabs have been used, install a new lock plate).

13 Fit the rubber boot over the lip on the left side of the swingarm, then install the shock absorber and the rear wheel.

7   Swingarm bearings – replacement

1 The swingarm bearings are press fit into the frame tube and must be driven out with a hammer and large punch. Insert the punch into the tube and drive the bearings out from the back side. Work carefully and do not gouge the inside of the frame tube.

2 The new bearings can be driven into the tube if it is done with extreme care, but the possibility of damaging the new bearings in the process is great. At any rate, the inside of the frame tube must be clean and free of burrs before the new bearings are installed. Lubricate the outside of the bearings and carefully drive them into place with a block of hardwood and a hammer until they are flush with or below the edge of the frame tube.

3 A preferred method would be to draw the bearings into place with a length of threaded rod and two nuts (or a large bolt and nut) and a couple of thick washers with a diameter slightly larger than the outside diameter of the bearing and frame tube. Slide one of the washers over the bolt or threaded rod with one nut installed, then slip a bearing over the bolt. The manufacturer's mark on the bearing should face out after it is installed. Insert the bolt through the frame tube, slip the remaining washer over the end and install the nut. Position the bearing in the frame tube opening and snug up the nut. Hold the bolt head with a wrench and tighten the nut until the bearing is drawn squarely into the tube. Disassemble the nut and bolt, then repeat the procedure to install the remaining bearing. Remember to lubricate the outside of the bearing and work carefully to avoid distorting it.

8   Fork oil – replacement

Refer to Chapter 1, *Tune-up and routine maintenance,* for the procedure to follow when replacing the fork oil.

9   Forks – removal and installation

1 The forks generally require very little maintenance as long as the fork oil level is maintained. In time, the seals will wear out and allow oil to accumulate on the sliders. When this occurs, the forks must be removed and new seals installed. **Note:** *To prolong seal life, keep the fork tubes between the sliders and the lower triple clamp clean.*

2 Remove the front wheel by referring to Chapter 7.

3 On RJ models, remove the brake caliper from the left fork slider (Chapter 7). Suspend the caliper from a piece of wire or an elastic cord; do not allow it to hang by the brake hose. RK models have two calipers.

4 Remove the front fender. It is held in place with four bolts.

5 Loosen both fork cap bolts (see Chapter 1, Section 35), then loosen the upper fork tube pinch bolts.

6 Remove the plastic horn cover. It is held in place by a bolt on each side (the bolts are hidden by plastic caps). Loosen the lower triple clamp bolts (photo) and carefully slide the fork tubes out of the clamps while twisting the tubes back and forth.

7 Installation is the reverse of removal. Be sure to tighten the fork cap bolts to the specified torque after tightening the lower triple clamp bolts. Also, do not overtighten the fender mounting bolts. **Note:** *On RK models, the air valves should face forward after the forks are installed.*

10   Forks – disassembly, inspection and reassembly

1 To properly disassemble and repair the front forks, you will need a bench vise, an eight (8) millimeter hex (allen) head socket, a 1/2-20 bolt and two nuts to fit the bolt, some new, clean fork oil of the recommended type and a clean place to work. *Always disassemble one fork leg at a time to avoid mixing up parts.*

2 Mount the fork tube in a vise. If the vise is not equipped with soft jaws, be sure to cushion the fork tube with rags or the exterior of the tube will be gouged and scratched. Remove the fork cap bolt, the spacer and the spring cap.

9.6   Loosen the lower triple clamp bolts (arrows) before removing the fork legs

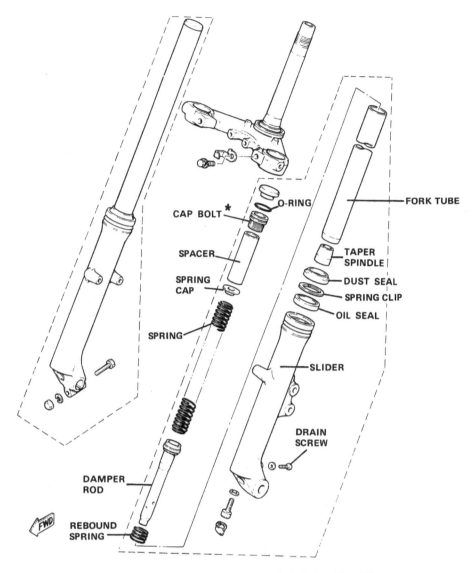

Fig. 6.6   Front fork components – exploded view (Sec 10)

* RK models have air valves threaded into the cap bolts, which look slightly different than the RJ model cap bolts

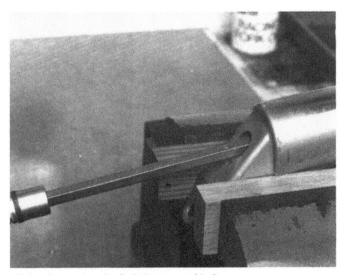

10.4a   Loosening the fork damper rod bolt

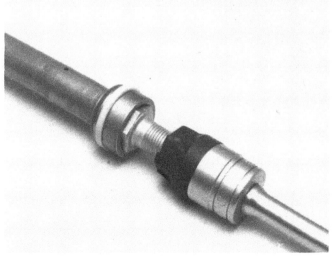

10.4b   Homemade damper rod holding tool (note how the bolt head fits into the upper end of the rod)

10.7a   Removing the dust seal with a screwdriver

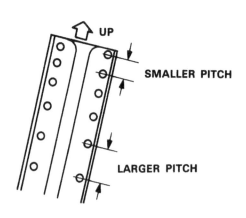

Fig. 6.7   The fork springs must be installed with the smaller pitch end up (Sec 10)

10.7b   Removing the fork seal spring clip

10.7c   Removing the fork seal (note the rag used to protect the slider)

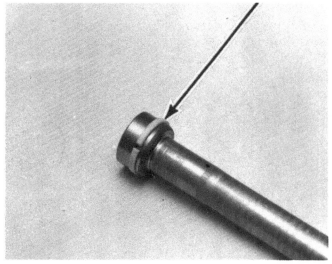

10.8a   Check the Teflon seal (arrow) on the damper rod for wear and damage (do not remove it unless a new one is to be installed)

10.8b   Check the slider bushing (arrow) for wear and damage

3   Remove the fork tube from the vise and drain the oil into a suitable container. Keep in mind that when the fork is turned upside down the spring will tend to slide out of place, so don't let it fall. Pump the fork tube and slider a few times to ensure complete draining. When the majority of the oil has drained, slide out the spring and wipe off as much of the oil as possible.

4   Next, carefully clamp the slider in the vise (be sure to cushion it) and remove the plastic plug and the eight (8) millimeter hex (Allen) head bolt from the bottom of the fork leg (photo). **Note:** *Clamp the slider at the axle mounting tab, not the hollow tube.* When the bolt is loosened, the damper rod inside the fork will probably turn also, preventing the bolt from loosening. To hold the damper rod, thread the two nuts onto the 1/2 – 20 bolt and tighten them against each other at the end of the bolt. Place the nuts into a 3/4 inch socket and wrap tape around the socket and bolt so they can't fall apart (the head of the bolt will be protruding). A long extension can be attached to the socket and the entire assembly can be lowered into the open end of the fork tube until the bolt head engages with the end of the damper rod (which has a recess shaped to accept the hex shape of the bolt head) (photo). A 1/2 – 20 bolt must be used because the damper rod recess is the same distance across as the distance across the flats of the bolt head (3/4 inch/19 mm). Before inserting the tool, compress the fork tube in the slider as far as possible. Attach a breaker bar to the extension and hold the damper rod as the bolt is loosened. The threads on the bolt should be cleaned thoroughly with a wire brush to remove any traces of thread locking compound (do not lose the copper sealing washer).

5   Withdraw the fork tube and damper rod assembly from the fork slider, then remove the damper rod and rebound spring from the fork tube. The taper spindle may come out with the damper rod assembly or it may stick in the bottom of the slider. Do not remove the Teflon ring from the damper rod. Use an appropriate size tap to remove any traces of thread locking compound from the threaded hole in the bottom of the damper rod.

6   Clean all components with solvent and dry them with compressed air.

7   Remove the dust seal and the fork seal spring clip from the slider, then pry out the seal with a screwdriver (photos). Cushion the slider with a rag so it doesn't get nicked.

8   Check the fork tube, the slider and the damper rod (photo) for score marks, scratches and excessive or abnormal wear. Look for dents in the slider and tube and check the slider bushing carefully for score marks (photo). Check the fork seal seat for nicks, gouges and scratches. If damage is evident, leaks will occur around the seal-to-slider junction. Replace any worn or defective parts with new ones.

9   Have the fork tube checked for runout at an automotive machine shop. **Warning:** *If it is bent, it should not be straightened; replace it with a new one.*

10   Measure the overall length of the spring and check it for cracks and other damage. If it is defective or sagged, replace both fork springs with new ones. Never replace only one spring.

11   Lubricate the outer circumference of the new seal with fork oil, then, using a large socket or similar tool and a hammer, install it in the slider. Apply pressure only on the outer edge of the seal and make sure it is seated squarely in the slider. Install the spring clip and make sure it is seated in its groove, then lubricate the seal lips with fork oil.

12   Place the rebound spring over the damper rod, slide the piston assembly into the fork tube until it protrudes from the lower end, then slip the taper spindle into place on the end of the damper rod. It may help to temporarily install the spring, spacer and fork cap bolt to keep pressure on the damper rod assembly. Be sure that the smaller pitch end of the spring is facing up.

13   Carefully slide the entire fork tube/damper rod assembly into the slider until the eight millimeter Allen head bolt can be threaded into the end of the fork damper rod.

14   Mount the slider in the vise, apply some liquid thread locking compound to the bolt, install the bolt (with sealing washer) through the slider and into the fork damper rod and tighten it to the specified torque. You will probably have to hold the damper rod with the homemade tool or it will turn and prevent the bolt from tightening.

15   Mount the fork tube in the vise with the open end up, then pour the specified amount of oil into the fork tube. A plastic baby bottle (calibrated in ounces and cc's) makes an ideal measuring device.

16   Insert the spring (if it is not already in place) with the smaller pitch end up. Lay the spring cap in place then install the spacer and cap bolt. This must be done against spring pressure, so be very careful not to damage the threads on the inside of the fork tube.

17   Store the completed fork leg in an upright position and repeat the rebuilding process for the remaining fork leg. Tighten the fork cap bolts after the forks have been reinstalled.

18   If the steering head bearings need maintenance or replacement, now is a very good time to do it since access to the bearings involves removal of the forks.

11   Steering head bearings — check and adjustment

Refer to Chapter 1 for this procedure.

12   Steering head bearings — maintenance

1   If the steering head bearing check/adjustment (Chapter 1) does not remedy excessive play or roughness in the steering head bearings, the entire front end must be disassembled and the bearings and races replaced with new ones.

2   Refer to Chapter 4 and remove the fuel tank. Refer to Section 9 and remove the front forks. Remove the headlight (Chapter 8), then refer to Section 3, disconnect the handlebars from the upper triple clamp and rotate them forward until they hang over the signal light stalks.

12.4   Removing the steering stem bolt

12.5   Loosening the upper ring nut with a spanner wrench

12.8   Check the bearing race in the steering head (arrow) for cracks, dents and pits

It is not necessary to remove the switches and throttle from the bars.
3   Remove the horn (it is held in place with one bolt), then remove the two bolts that attach the headlight bracket to the lower triple clamp. Pull down on the headlight assembly to disengage the bracket from the upper triple clamp, then let the headlight/signal light assembly and the handlebars rotate forward. Remove the instrument cluster (two nuts), the key switch (two bolts) and the warning light panel (held in place by two tabs), then rotate all of the components that were attached to the triple clamps forward and down and support them on a table, a trash can or a barrel (covered with an old blanket or towel).
4   Remove the steering stem bolt (photo), then lift off the upper triple clamp (sometimes called the fork bridge or crown).
5   Using a pin-type spanner wrench, remove the ring nuts (photo) while supporting the steering head from the bottom. Lift off the race cover.
6   Place a clean drain pan directly under the steering head and carefully withdraw the stem and lower triple clamp. Hopefully, any ball bearings that fall out of place will land in the drain pan. Lift the upper ball race off and remove the steel balls from the frame steering head.
7   Clean all the parts with solvent and dry them thoroughly. Wipe the old grease out of the frame steering head and ball races.
8   Examine the races in the frame steering head (photo) for cracks, dents, and pits. If even the slightest amount of wear or damage is evident, the races should be replaced with new ones.
9   To remove the races, drive them out of the steering head with a

12.9   Removing the lower race with a drift punch and hammer

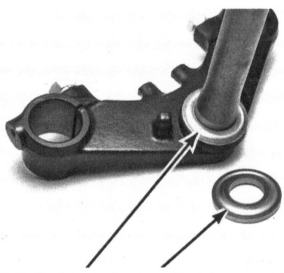

12.10   Check the steering stem and top ball race as well (arrows)

12.13   After packing the steering stem race with grease, lay 19 of the bearings in place

12.14   The remaining 19 bearings are installed in the top frame race (use grease to hold them in place)

large drift punch (held against the back side of the race) and a hammer (photo). Since the races are an interference fit in the frame, installation will be easier if the new races are left overnight in a refrigerator. this will cause them to contract and slip into place in the frame with very little effort. When installing the races, tap them gently into place with a hammer and punch or a large socket. *Do not strike the bearing surface or the race will be damaged.*

10   Check the steering stem race, the top race (photo) and the ball bearings. Look for cracks, dents, and pits in the races and flat spots on the bearings. Replace any defective parts with new ones. If any new races or bearings are required, replace all of the races and all of the bearings with new parts.

11   To remove the race from the steering stem, simply tap evenly around its outer circumference with a hammer and punch. Reverse the pro-

cess to install the new race and be very careful not to strike the bearing surface.

12   Inspect the steering stem, the triple clamp and the fork bridge for cracks and other damage. Do not attempt to repair any steering components. Replace them with new parts if defects are found.

13   Pack the steering stem race with grease (preferably one containing molybdenum disulfide), then lay 19 of the bearings in place in the race (photo). The grease should hold them securely in place during installation of the steering stem.

14   Pack the upper race in the frame steering head with grease, then lay 19 of the bearings in place on the race (photo).

15   Pack the lower race (in the frame steering head) and the top race with grease. Carefully slide the steering stem into place in the frame, slip the top race over the stem and install the race cover and the ring

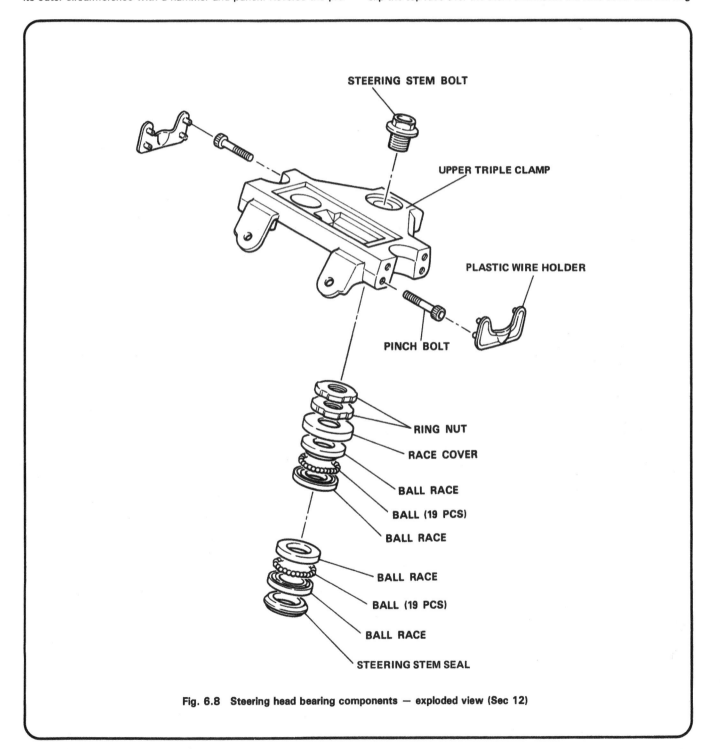

Fig. 6.8 Steering head bearing components — exploded view (Sec 12)

nuts. Tighten the lower ring nut until it is snug, then back it off slightly. Make sure the stem rotates freely with no noticeable vertical play.
16 Hold the lower ring nut with a spanner wrench and tighten the upper ring nut against it. Install the upper triple clamp and the steering stem bolt. Tighten the bolt to the specified torque while holding the upper triple clamp in alignment (it may help to slip one of the fork legs into place temporarily to keep the upper clamp in alignment).
17 Check the steering head bearings again for excessive play and binding. The steering stem should rotate freely without any noticeable vertical play. If readjustment is necessary, loosen the steering stem bolt, turn the ring nuts as necessary, then retighten the stem bolt to the specified torque.
18 Reinstall the warning light panel, the key switch, the instrument cluster, the headlight/signal light assembly, the handlebars, the forks, the front wheel and the fuel tank.

13 Side and centerstand – maintenance

1 The centerstand pivots on two bolts attached to the frame. Periodically, remove the pivot bolts and grease them thoroughly to avoid excessive wear.
2 Make sure that the return spring is in good condition. A broken or weak spring is an obvious safety hazard.
3 The sidestand is attached to a bracket bolted to the engine and frame. An extension spring anchored to the bracket ensures that the stand is held in the retracted position.
4 Make sure that the pivot bolt is tight and that the extension spring is in good condition and not overstretched. An accident is almost certain to occur if the stand extends while the machine is in motion.

14 Final drive gear oil level – check

Refer to Chapter 1, *Tune-up and routine maintenance*, for the gear oil level checking procedure.

15 Final drive gear oil – change

Refer to Chapter 1 for the procedure to follow when changing the final drive gear oil.

16 Shaft drive assembly – general information

The shaft drive assembly consists of the output gear assembly, which is attached to the engine, the driveshaft and U-joint, which run inside the lower left swingarm tube, and the final drive assembly, which is attached to the rear of the left swingarm tube and mates with the rear wheel hub.

Because of the special tools and expertise required to diagnose, service and repair the shaft drive components, it is recommended that any work required be left to a Yamaha dealer service department. Due to the fact that most of the components run in an oil bath, problems with this type of a drivetrain are few and far between.

17 Suspension – adjustment (RK models)

Front forks

1 Place the motorcycle on the centerstand.
2 Raise the front wheel so no weight is on the front end of the motorcycle.
3 Remove the dust cap from the air valve on top of each fork leg.
4 Using an air gauge, check the air pressure.
5 If the air pressure is decreased, the suspension becomes softer. If air pressure is increased, the suspension becomes harder. The air pressure should always remain between 6 and 12 psi. **Note:** *A manual air pump is the easiest way to add a controlled amount of air.*

Rear shock

Spring preload
6 The rear shock spring preload can be adjusted for the rider's weight, ride preference and road conditions.
7 Open the seat.
8 If the preload is to be increased, use the shock nut wrench included in the motorcycle tool kit and raise the spring seat by turning it.
9 If the preload is to be decreased, use the shock nut wrench and lower the spring seat.
10 Close the seat.

Dampening
11 Remove the plastic side cover from the right side of the motorcycle.
12 Turning the dampening adjuster nut will increase or decrease the shock dampening as desired. **Note:** *If the adjuster is turned toward the no. 1, the dampening becomes softer. If the adjuster is turned toward the no. 5, the dampening becomes harder. Position no. 3 is the standard position.*

Chapter 7 Wheels, brakes and tires

Contents

Specifications

Wheels and tires

Wheel runout limit	
Axial	0.079 in (2 mm)
Radial	0.0079 in (0.2 mm)
Minimum tire tread depth	See Chapter 1
Tire pressures	See Chapter 1
Tire sizes – UK model	
Front	90/90-18 51H Tubed
Rear	4.25/85 H18 Tubed
Tire sizes – US model	
Front	90/90-18 51H Tubeless
Rear	110/90-18 61H Tubeless

Front disc brake

Front brake lever free play	See Chapter 1
Brake pad wear limit	0.031 in (0.8 mm)
Brake disc runout limit	0.006 in (0.15 mm)
Brake disc minimum thickness	0.180 in (4.5 mm)
Brake fluid type	DOT 3

Rear drum brake

Rear brake pedal free play	See Chapter 1
Rear brake shoe lining wear limit	5/64 in/0.080 in (2 mm)
Rear brake drum inside diameter	
Standard	7.09 in (180 mm)
Service limit	7.13 in (181 mm)

Torque specifications	Ft-lb	Nm
Front axle nut	80	110
Front axle pinch bolt/nut	14	20
Rear axle nut	80	110
Rear axle pinch bolt	4.3	6
Disc brake pad retaining bolt	17	23
Disc brake caliper mounting bolts	25	35
Disc brake caliper guide bolts	19	26
Brake hose banjo fitting union bolts	14.5	20
Brake disc mounting bolts	14.5	20
Final drive hub mounting bolts	29	40

1   General information

The Yamaha Vision is equipped with cast aluminum wheels, which require very little maintenance and allow tubeless tires to be used. Brakes are hydraulic disc on the front (RK and European models are equipped with dual discs) and single leading shoe drum at the rear.

Caution: *Disc brake components rarely require disassembly. Do not disassemble components unless absolutely necessary. If any hydraulic brake line connection in the system is loosened, the entire system should be disassembled, drained, cleaned and then properly filled and bled upon reassembly. Do not use solvents on internal brake components. Solvents will cause seals to swell and distort. Use only clean brake fluid or alcohol for cleaning. Use care when working with brake fluid as it can injure your eyes and it will damage painted surfaces and plastic parts.*

2   Brake shoes/pads — wear check

Since this procedure is considered part of routine maintenance, refer to Chapter 1.

3   Brake system — general check

Refer to Chapter 1, *Tune-up and routine maintenance*, for the brake system general check.

4   Brake light switches — check and adjustment

The brake light switch check and adjustment procedures (front and rear) are included in Chapter 1.

5   Rear brake pedal/front brake lever — check and adjustment

The brake pedal and lever adjustments are considered part of routine maintenance and are included in Chapter 1.

6   Tires/wheels — general check

Refer to Chapter 1, *Tune-up and routine maintenance.*

7   Wheels — inspection and repair

1   Place the motorcycle on the centerstand, then clean the wheels thoroughly to remove mud and dirt that may interfere with the inspection procedure or mask defects. Make a general check of the wheels and tires as described in Chapter 1, *Tune-up and routine maintenance.*
2   With the motorcycle on the centerstand and the front wheel in the air, attach a dial indicator to the fork slider and position the stem against the side of the rim (photo). Spin the wheel slowly and check the side-to-side (axial) runout of the rim. In order to accurately check radial runout with the dial indicator, the wheel would have to be removed from the machine and the tire removed from the wheel. With the axle clamped in a vise, the wheel can be rotated to check the runout.
3   An easier, though slightly less accurate, method is to attach a stiff wire pointer to the fork slider and position the end a fraction of an inch from the wheel (where the wheel and tire join). If the wheel is true, the distance from the pointer to the rim will be constant as the wheel is rotated. Repeat the procedure to check the runout of the rear wheel.
Note: *If wheel runout is excessive, refer to the appropriate Section in this Chapter and check the wheel bearings very carefully before replacing the wheel.*
4   The wheels should also be visually inspected for cracks, flat spots on the rim and other damage. Since tubeless tires are involved, look very closely for dents in the area where the tire bead contacts the rim.

Dents in this area may prevent complete sealing of the tire against the rim, which leads to deflation of the tire over a period of time.
5   If damage is evident, or if runout in either direction is excessive, the wheel will have to be replaced with a new one. *Never attempt to repair a damaged cast aluminum wheel.*

8   Wheels — alignment check

1   Misalignment of the wheels, which may be due to a cocked rear wheel or a bent frame or triple clamps, can cause strange and possibly serious handling problems. If the frame or triple clamps are at fault, repair by a frame specialist or replacement with new parts are the only alternatives.
2   To check the alignment you will need an assistant, a length of string or a perfectly straight piece of wood and a ruler graduated in 1/64 inch increments. A plumb bob or other suitable weight will also be required.
3   Place the motorcycle on the centerstand, then measure the width of both tires at their widest points. Subtract the smaller measurement from the larger measurement, then divide the difference by two. The result is the amount of offset that should exist between the front and rear tires on both sides.
4   If a string is used, have your assistant hold one end of it about half way between the floor and the rear axle, touching the rear sidewall of the tire.
5   Run the other end of the string forward and pull it tight so that it is roughly parallel to the floor. Slowly bring the string into contact with the front sidewall of the rear tire, then turn the front wheel until it is parallel with the string. Measure the distance from the front tire sidewall to the string.
6   Repeat the procedure on the other side of the motorcycle. The distance from the front tire sidewall to the string should be equal on both sides.
7   As was previously pointed out, a perfectly straight length of wood may be substituted for the string. The procedure is the same.
8   If the distance between the string and tire is greater on one side, or if the rear wheel appears to be cocked, refer to Chapter 6, *Swingarm bearings — check,* and make sure the swingarm is tight.
9   If the front-to-back alignment is correct, the wheels still may be out of alignment vertically.
10   Using the plumb bob, or other suitable weight, and a length of string, check the rear wheel to make sure it is vertical. To do this, hold the string against the upper tire sidewall and allow the weight to settle just off the floor. When the string touches both the upper and lower tire sidewalls and is perfectly straight, the wheel is vertical. If it is not, place thin spacers under one leg of the centerstand.
11   Once the rear wheel is vertical, check the front wheel in the same manner. If both wheels are not perfectly vertical, the frame and/or major suspension components are bent.

7.2   Checking axial runout of the wheel rim with a dial indicator

9   Front wheel – removal and installation

1   Place the motorcycle on the centerstand, then raise the front wheel off the ground by placing a block or jackstand under the engine.
2   It is not necessary, but you may want to disconnect the spedometer cable from the drive unit.

3   Remove the cotter key, unscrew the axle nut (photo) and remove the large washer. Remove the pinch bolt from the right fork leg (photo).
4   Support the wheel, remove the axle (photo) and carefully lower the wheel until the brake disc clears the caliper. On RK and European models (dual discs), lower the wheel until the discs clear the calipers, then rotate the fork sliders to move the calipers out and allow room for the tire to pass between them. Do not lose the spacer that fits into the right side of the hub. **Note:** *Do not operate the front brake lever*

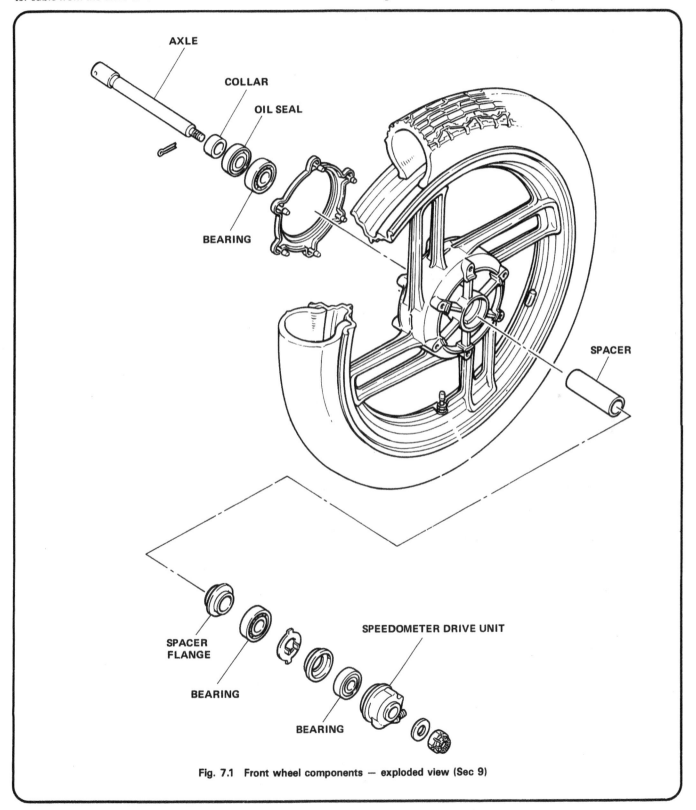

Fig. 7.1 Front wheel components — exploded view (Sec 9)

with the wheel removed. If the axle is corroded, remove the corrosion with fine emery paper.

5   Installation is the reverse of removal. Apply a thin coat of grease to the seal lip, then slide the collar into the right side of the hub (photo). Position the speedometer drive unit in place in the left side of the hub, then slide the wheel into place. Make sure the notch in the speedometer drive housing lines up with the lug on the left fork leg. If the disc will not slide between the brake pads, remove the wheel and carefully pry them apart with a piece of wood.

6   Slip the axle into place, then tighten the axle nut to the specified torque. The axle can be kept from turning by placing a screwdriver through the hole in the right end. After the axle nut has been tightened, apply the front brake, pump the forks up and down several times and check for binding.

7   Install the pinch bolt, lockwasher and nut and tighten it to the specified torque, then install a new cotter key in the left end of the axle.

10   Front disc brake — inspection

1   Carefully examine the master cylinder, the hose and the caliper unit for evidence of brake fluid leakage. Pay particular attention to the hose. If it is cracked, abraded, or otherwise damaged, replace it with a new one. If leaks are evident at the master cylinder or caliper,

they should be rebuilt by referring to the appropriate Sections in this Chapter.

2   Check the lever for proper operation. It should feel firm and return to its original position when released. If it feels spongy, or if lever travel is excessive, the system may have air trapped in it. Refer to Section 16 and bleed the brakes.

3   Check the brake pads for excessive wear by referring to Chapter 1, *Tune-up and routine maintenance.*

4   Examine the brake disc for cracks and evidence of scoring. Measure the thickness of the disc (photo) and compare it to the Specifications. If it has worn beyond the allowable limit, it must be replaced with a new one.

5   If the brake lever pulsates when the brake is applied during operation of the machine, the disc may be warped. Attach a dial indicator set-up to the fork slider and check the disc runout (photo). If the runout is greater than specified, replace the disc with a new one. If a dial indicator is not available, a dealer service department or motorcycle repair shop can make this check for you.

11   Front disc brake — pad replacement

1   If the brake pads are worn out (refer to Chapter 1) or contaminated with brake fluid or dirt, they must be replaced with new ones. Failure

9.3a   Removing the front axle nut

9.3b   Removing the front axle pinch bolt and nut

9.4   Insert a Phillips screwdriver through the hole in the axle, then twist and pull the axle out of the forks

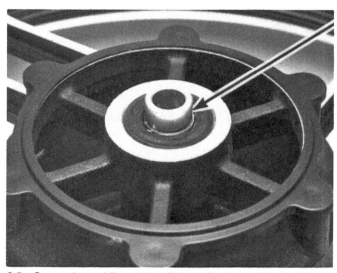

9.5   Grease the seal lip and install the collar (arrow) in the right side of the hub before attaching the wheel to the machine

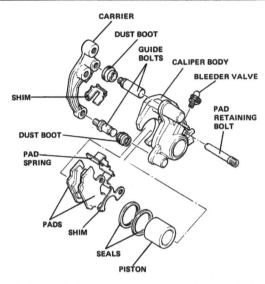

Fig. 7.2 Disc brake caliper components — exploded view (Sec 11)

to replace the pads when necessary will result in damage to the disc and severe loss of stopping power. **Note:** *Always replace the brake pads as a set.*

2   To remove the pads, simply unscrew the pad retaining bolt from the caliper with a six (6) millimeter hex (Allen) head wrench (photo) and withdraw the pads and shim from the bottom of the caliper. Note how the shim is positioned as the pads are withdrawn. Once the pads have been removed from the caliper, it will be much easier to tell to what extent they have worn. If in doubt, measure the thickness of the remaining lining material and compare it to the Specifications.

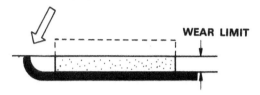

Fig. 7.3 When measuring the disc brake pad lining material, do not include the metal backing plate (Sec 11)

10.4   Measuring the brake disc thickness with a micrometer

10.5   Checking the brake disc runout with a dial indicator

11.2   Removing the disc brake pad retaining bolt from the caliper

11.4   Proper installed position of the pad shim (inner pad)

3   Clean the disc surface and the lining of the new pads with brake system cleaner (available at auto parts stores), lacquer thinner or acetone. *Do not use petroleum-based solvents.*
4   Do not touch the lining surface when installing the new pads. Make sure the new shim is positioned exactly as the old one was (photo) as the pads were removed, then slip the new outer pad into the caliper (you may have to pull out on the caliper to make room for the new thicker pad). Push the retaining bolt part way in to support the outer pad (make sure it slips through the hole in the pad tab as it is installed). Push in on the caliper, then slip the inner pad into place and push the retaining bolt in until the threads can engage the caliper (make sure it slips through the hole in the pad).
5   Tighten the retaining bolt to the specified torque.
6   Check the brake fluid level by referring to Chapter 1. After installing new brake pads, the level may be too high. If so, remove the reservoir cover and carefully siphon off the excess fluid.

12   Front brake disc — removal and installation

1   The brake disc is attached to the front hub with six bolts and is easily removed.
2   Remove the front wheel by referring to Section 9, then bend back the locking tabs and unbolt the disc (photo).
3   Before installing the disc, thoroughly clean the bolt threads and the threaded holes in the hub. Use a thread locking compound and be sure to tighten the bolts evenly and gradually, in a criss-cross pattern, until the specified torque is reached.
4   Bend up the *unused* locking tabs to prevent the bolts from loosening. If both of the locking tabs at each bolt head have been used, the lock plates must be replaced with new ones.

13   Front disc brake caliper — removal and installation

1   Before removing the caliper, obtain an 8 x 20 millimeter bolt and nut (preferably a locknut with a nylon insert).
2   Loosen the caliper banjo fitting bolt (photo) and pull the hose and bolt away from the caliper as a unit. While holding the bolt in the hose fitting, quickly place your finger over the end of the hollow bolt to prevent loss of brake fluid.
3   The eight (8) millimeter bolt mentioned previously, when used to replace the hollow banjo fitting bolt, will effectively seal off the hydraulic system and make the job of bleeding the brakes much easier. Working quickly, remove the hollow bolt, slip the solid bolt (along with the sealing washers) through the banjo fitting, thread on the nut (photo) and tighten it securely. Wrap the end of the hose in a rag, as some fluid will seep through. Wipe up any spilled brake fluid.
4   Remove the brake pads as described in Section 11.
5   Remove the caliper mounting bolts (photo) and separate the caliper

12.2   Bend back the locking tabs (arrow) before removing the disc mounting bolts

13.2   Loosening the brake caliper banjo fitting bolt

13.3   Plug the hose banjo fitting with a nut and bolt to prevent excessive loss of brake fluid

13.5   Removing the caliper mounting bolts

from the fork slider.

6   Installation is the reverse of removal. Use thread locking compound on the caliper mounting bolts and tighten them to the specified torque. The caliper assembly must be free to slide back and forth on the carrier. If there is any sign of binding, or if the caliper assembly is cocked, remove the caliper and refer to Section 14 for the guide bolt inspection procedure. Install the pads, hook up the brake hose and bleed the system as described in Section 16.

14   Front disc brake caliper — overhaul

1   If the caliper is leaking fluid around the piston, it should be removed and overhauled to restore braking performance. Before disassembling the caliper, read through the entire procedure and make sure you have the correct caliper rebuild kit. Also, you will need some new, clean brake fluid of the recommended type, some clean rags and a clean place to work.

2   **Caution:** *Disassembly, overhaul and reassembly of the brake caliper must be done in a spotlessly clean work area to avoid contamination and possible failure of the brake hydraulic system components. If such a work area is not available, have the caliper rebuilt by a dealer service department or a motorcycle repair shop.*

3   Remove the caliper as described in Section 13.

4   Carefully separate the carrier from the caliper body. Detach the pad spring and the rubber dust boots from the caliper body and separate the shim from the carrier. **Note:** *Do not remove the guide bolts from the carrier unless they must be replaced with new ones.*

5   Grasp the piston by hand and pull it straight out of the caliper. Do not use pliers and do not pry on the piston with metal tools, as it may be damaged. If the piston is difficult to remove, reconnect the brake hose to the caliper body, position the caliper over a box stuffed with clean rags and slowly depress the brake lever. The hydraulic pressure will force the piston out of the caliper. Do not allow the piston to fall onto a hard surface, as damage will result. If compressed air is available, it can also be used to remove the piston. The pressure must be kept extremely low or the piston can fly out of the caliper and be damaged or cause injury. Place the air nozzle close to the hose fitting (do not hold it against the hole) and position the caliper over a box stuffed with clean rags.

6   Remove the rubber seals. Carefully push them into the bore (photo), then pull them out. To avoid scratching the caliper bore, use a wood or brass tool to remove the seals.

7   Clean all of the brake components (except for the brake pads) with brake cleaning solvent (available at auto part stores), isopropyl alcohol or clean brake fluid. **Caution:** *Do not, under any circumstances, use a petroleum-based solvent to clean brake parts.* If compressed air is available, use it to dry the parts thoroughly.

8   Check the caliper bore and the outside of the piston (photo) for corrosion, scratches, nicks and score marks. If damage is evident, the

14.6   To remove the rubber seal, first push it into the bore (do not use a steel tool), then pull it out

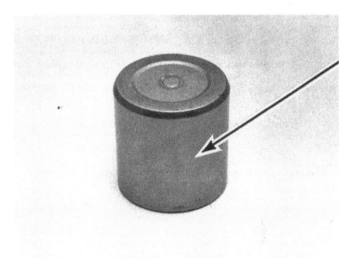

14.8   Check the caliper piston carefully for score marks and scratches in the area indicated by the arrow

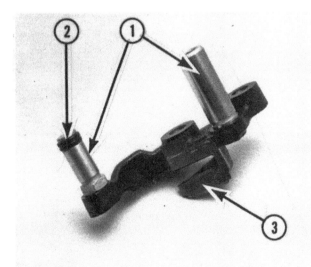

14.9   Caliper carrier components; guide bolts (1), O-ring (2) and shim (3)

14.13   Attach the rubber dust boots (1) and the pad spring (2) to the caliper as shown

caliper must be replaced with a new one. **Note:** *If the caliper is in poor condition, then the master cylinder should be checked as well.*

9   Check the caliper holes, the carrier and the guide bolts for damage, corrosion and evidence of excessive wear (photo). If the rubber dust boots are cracked or otherwise damaged, replace them with new ones.

10   Remove the O-ring from the short guide bolt and replace it with a new one. Before slipping it over the bolt, lubricate it with silicone grease. Also, apply a thin coat of grease to the sliding portions of the guide bolts.

11   Lubricate the new seals with clean brake fluid and install them in the grooves in the caliper bore. Make sure the seal with the square cross-section is installed in the inner groove.

12   Apply a coat of clean brake fluid to the inside of the caliper bore, then carefully push the piston into the caliper with the open side out.

13   Attach the rubber dust boots and the pad spring to the caliper (photo), then position the shim on the carrier

14   Attach the caliper to the carrier. Make sure the rubber dust boots are properly seated in the guide bolt and caliper grooves.

15   Refer to the appropriate Sections and attach the caliper to the fork slider, reinstall the brake pads and bleed the system.

---

15   Front disc brake master cylinder — removal, overhaul and installation

1   If the master cylinder is leaking fluid, or if the lever does not produce a firm feel when the brake is applied, and bleeding the brakes does not help, master cylinder overhaul is recommended. Before disassembling the master cylinder, read through the entire procedure and make sure that you have the correct rebuild kit. Also, you will need some new, clean brake fluid of the recommended type, some clean rags and internal snap-ring pliers. **Note:** *To prevent damage to the paint from spilled brake fluid, always cover the gas tank when working on the master cylinder.*

2   **Caution:** *Disassembly, overhaul and reassembly of the brake master cylinder must be done in a spotlessly clean work area to avoid contamination and possible failure of the brake hydraulic system components. If such a work area is not available, have the master cylinder rebuilt by a dealer service department or motorcycle repair shop.*

Removal

3   Remove the rear view mirror and loosen, but do not remove, the screws holding the reservoir top cover in place.

4   Pull back the rubber boot, loosen the banjo fitting bolt (photo) and separate the brake hose from the master cylinder. Wrap the end of the hose in a clean rag and suspend the hose in an upright position or bend it down carefully and place the open end in a clean container. The objective is to prevent excess loss of brake fluid and fluid spills.

5   Insert a small screwdriver into the hole in the underside of the lever mount, depress the tab (photo) and separate the front brake light switch

from the mount by pulling straight out on it.

6   Remove the locknut from the underside of the lever pivot bolt, then unscrew the bolt (photo).

7   Remove the master cylinder mounting bolts (photo) and separate the master cylinder from the handlebar. **Caution:** *Do not tip the master cylinder upside down or brake fluid will run out.*

Overhaul

8   Detach the top cover and the rubber diaphragm, then drain the brake fluid into a suitable container. Wipe any remaining fluid out of the reservoir with a clean rag.

9   Carefully remove the rubber dust boot from the piston assembly (photo).

10   Using snap-ring pliers, remove the snap-ring (photo) and slide out the piston, the cup seals and the spring. Lay the parts out in the proper order to prevent confusion during reassembly.

11   Clean all of the parts with brake cleaning solvent (available at auto parts stores), isopropyl alcohol or clean brake fluid. **Caution:** *Do not, under any circumstances, use a petroleum based solvent to clean brake parts.* If compressed air is available, use it to dry the parts thoroughly. Check the master cylinder bore for corrosion, scratches, nicks and score marks. If damage is evident, the master cylinder must be replaced with a new one. *If the master cylinder is in poor condition, then the caliper should be checked as well.*

15.4   Loosening the master cylinder banjo fitting bolt

15.5   Removing the brake light switch

15.6   Removing the brake lever pivot bolt

12 Remove the old cup seals from the piston and install the new ones. Make sure the lips face away from the lever end of the piston (photo). If a new piston is included in the rebuild kit, use it regardless of the condition of the old one.
13 Before reassembling the master cylinder, soak the piston and the rubber cup seals in clean brake fluid for ten or fifteen minutes. Lubricate the master cylinder bore with clean brake fluid, then carefully insert the piston and related parts in the reverse order of disassembly. Place the spring on the piston and slip them into the bore together. Make sure the lips on the cup seals do not turn inside out when they are slipped into the bore.
14 Depress the piston, then install the snap-ring (make sure the snap-ring is properly seated in the groove with the sharp edge facing out). Install the rubber dust boot (make sure the lip is seated properly in the piston groove).

Installation

15 Attach the master cylinder to the handlebar and tighten the bolts securely. The arrow and the word 'up' on the master cylinder clamp should be pointing up and readable. Install the brake lever and tighten the pivot bolt locknut.
16 Install the brake light switch, connect the brake hose to the master cylinder and install the mirror. Refer to Section 16 and bleed the air from the system, then install the master cylinder diaphragm and cover.

16 Front disc brake — bleeding procedure

1 Bleeding the brake is simply the process of removing all the air bubbles from the brake fluid reservoir, the lines and the brake caliper. Bleeding is necessary whenever a brake system hydraulic connection is loosened, when a component or hose is replaced, or when the master cylinder or caliper is overhauled. Leaks in the system may also allow air to enter, but leaking brake fluid will reveal their presence and warn you of the need for repair.
2 To bleed the brake, you will need some new, clean brake fluid of the recommended type, a length of clear vinyl or plastic tubing, a plastic container, some rags and a wrench to fit the brake caliper bleeder valve.
3 Cover the gas tank and other painted components to prevent damage in the event that brake fluid is spilled.
4 Attach one end of the clear vinyl or plastic tubing to the brake caliper bleeder valve and submerge the other end in some clean brake fluid in the plastic container.
5 Remove the reservoir cover and check the fluid level. Do not allow the fluid level to drop below the lower mark during the bleeding process.
6 Carefully pump the brake lever three or four times and hold it while opening the caliper bleeder valve (photo). When the valve is opened, brake fluid will flow out of the caliper into the clear tubing

15.7 Removing the master cylinder mounting bolts

15.9 The rubber dust boot (arrow) is seated in a groove in the end of the piston (do not tear it during removal)

15.10 Remove the snap-ring (arrow) and withdraw the piston

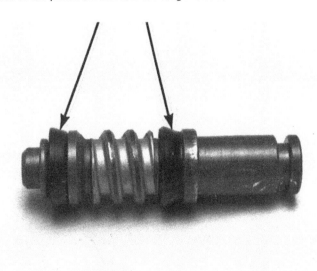

15.12 Install the cup seals (arrows) with the lips facing away from the lever end of the piston

and the lever will move toward the handlebar.

7   Retighten the bleeder valve, then release the brake lever gradually. Repeat the process until no air bubbles are visible in the brake fluid leaving the caliper and the lever is firm when applied. Remember to add fluid to the reservoir as the level drops. Use only new, clean brake fluid of the recommended type. Never reuse the fluid lost during bleeding.

8   Replace the reservoir cover, wipe up any spilled brake fluid and check the entire system for leaks. **Note:** *If bleeding is difficult, it may be necessary to let the brake fluid in the system stabilize for a few hours. Repeat the bleeding procedure when the tiny bubbles in the system have settled out.*

17   Rear wheel — removal and installation

1   Place the motorcycle on the centerstand.

2   Remove the cotter key, nut and washers, then separate the brake tension bar from the backing plate.

3   Unscrew the brake rod adjusting nut, depress the brake pedal and separate the brake rod from the brake arm. Do not lose the nut, the spring or the brake rod fitting.

4   Pull out the cotter key and remove the axle nut, then loosen the right side pinch bolt (photo). Withdraw the axle from the right side

and catch the spacer. You may have to tap lightly on the threaded end with a soft-faced hammer to dislodge the axle.

5   Move the wheel (with the brake in place) to the right until it is detached from the final drive assembly, then withdraw it to the rear. Do not allow the brake to fall out of the hub.

6   To install the wheel, make sure the collar is in place in the final drive assembly and apply a light coat of grease to the large O-ring (photos), then insert the brake assembly into the wheel hub. Line up the splines in the final drive assembly and the hub and slide the wheel into place.

7   Slip the axle through the swingarm and the spacer, then push it through the wheel hub and final drive unit.

8   Install the axle nut and tighten it to the specified torque, then install a *new* cotter key. Tighten the pinch bolt on the right side of the swingarm to the specified torque.

9   Attach the brake rod to the brake arm, then hook up the brake tension bar, tighten the nut and install a *new* cotter key.

10   Refer to Chapter 1 and adjust the rear brake pedal free play.

18   Rear brake — inspection and brake shoe replacement

1   Drum brakes do not usually require frequent maintenance, but they should be checked periodically to ensure proper operation. If the linkage

16.6   Opening the brake caliper bleeder valve (note how the plastic tube is attached)

17.4   Loosening the rear axle pinch bolt

17.6a   Make sure the collar (arrow) is in place before installing the rear wheel

17.6b   Apply grease to the clutch hub O-ring seal (arrow) before installing the rear wheel

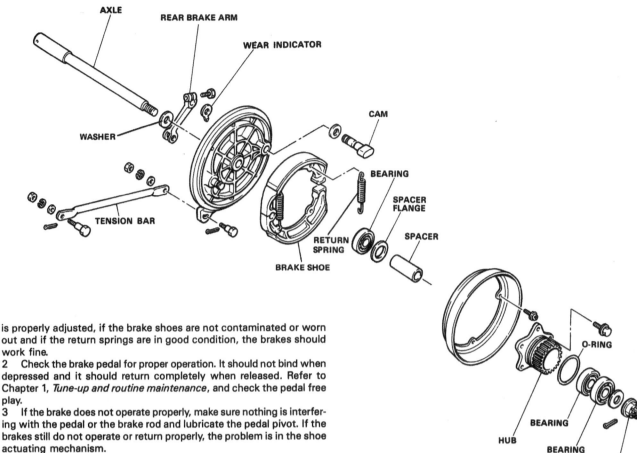

is properly adjusted, if the brake shoes are not contaminated or worn out and if the return springs are in good condition, the brakes should work fine.

2   Check the brake pedal for proper operation. It should not bind when depressed and it should return completely when released. Refer to Chapter 1, *Tune-up and routine maintenance*, and check the pedal free play.

3   If the brake does not operate properly, make sure nothing is interfering with the pedal or the brake rod and lubricate the pedal pivot. If the brakes still do not operate or return properly, the problem is in the shoe actuating mechanism.

4   If the brake shoe wear check (refer to Chapter 1, *Tune-up and routine maintenance*) indicates that the shoes are near the wear limit, refer to Section 17 and remove the rear wheel. Measure the thickness of the brake shoe lining (photo) and compare it to the Specifications. If the shoes have worn beyond the allowable limits, or if they are worn unevenly, they must be replaced with new ones.

5   If the linings are acceptable as far as thickness is concerned, check them for glazing, high spots and hard areas. A light touch-up with a file or emery paper will restore them to usable condition. If the linings are extremely glazed, they have probably been dragging. Be sure to properly adjust the pedal free play to prevent further glazing.

6   Occasionally the linings may become contaminated with grease from the wheel bearing or brake cam. If this happens, and it is not too severe, cleaning the shoes with a brake system solvent (available at auto parts stores) may restore them. Better yet, replace the shoes with new ones.

Fig. 7.4   Rear wheel hub and drum brake components — exploded view (Sec 18)

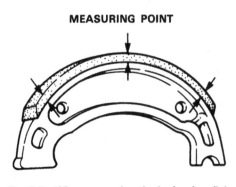

Fig. 7.5   When measuring the brake shoe lining thickness, look for uneven wear by taking the measurements at three places (Sec 18)

18.4   Measuring the drum brake shoe lining thickness

7   To remove the shoes from the backing plate, pull up on the outer edges of the shoes until the shoes form a 'V', then lift them away from the backing plate. Remove the springs from the shoes and check them for cracks and excessive stretch. Replace them with new ones if defects are noted.

8   Using a center punch, put alignment marks on the brake arm and shaft (to simplify reassembly), then remove the brake arm from the shaft by unscrewing the bolt from the arm. Lift off the brake wear indicator, then slip the shaft and washer out of the backing plate. Clean the backing plate components with solvent to remove brake dust and dirt. **Caution:** *Do not use compressed air to blow the dust off the backing plate — it can be hazardous if inhaled.* If compressed air is available, use it to dry the parts thoroughly.

9   Check the shaft and the hole in the backing plate for signs of excessive wear. Slide the shaft back into the backing plate and make sure it turns smoothly without binding. If excessive side play is evident, the backing plate will have to be replaced with a new one. Also, check the shoe contact areas of the cam for wear (photo).

10   Apply a thin coat of high-temperature grease to the cam shaft and install it in the backing plate (don't forget the washer that fits under the cam). Slip the wear indicator (photo) over the shaft. Align the punch marks on the shaft and brake arm, slide the brake arm onto the shaft, install the bolt and tighten it securely.

11   Before installing the new shoes, file a taper on their leading edges.

Install the springs, then apply a thin coat of high-temperature grease to the shoe contact areas of the cam and the pivot (photo). Hold the shoes in a 'V' and set them in position, then push down on them until they are properly engaged with the cam and pivot.

12   Using a stiff brush, remove the dust from the drum. **Caution:** *Do not, under any circumstances, use a petroleum based solvent to clean the drum.*

13   Check the drum for rough spots, rust and evidence of excessive wear. If the outer edge of the drum has a pronounced ridge, excessive wear has occurred. To confirm this, measure the drum diameter (photo) and compare it to the Specifications. Make the measurement at several places to determine if the drum is out-of-round. Excessive wear and out-of-roundness indicate the need for a new hub/drum. Slight roughness and rust spots can be removed with fine emery paper. Use one of the brake shoes as a sanding block so low spots aren't created in the drum.

14   Refer to Section 17 and install the rear wheel on the machine.

19   Wheel bearings — inspection and maintenance

1   Wheel bearing maintenance is often neglected because the bearings are relatively inaccessible. The bearings should be removed, cleaned, inspected and repacked at 25 000 mile intervals (more often if the

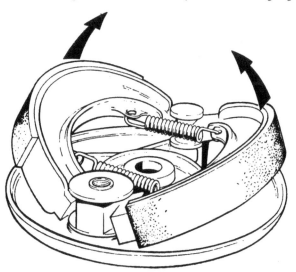

Fig. 7.6   Removing the brake shoes from the backing plate — reverse the procedure to install them (Sec 18)

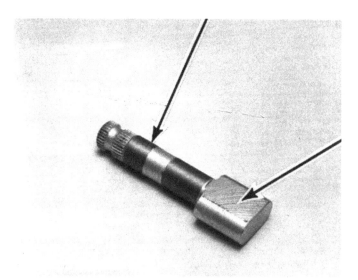

18.9   Check the brake cam shaft and the shoe contact areas (arrows) for wear

18.10   The large tooth on the wear indicator (arrow) must be aligned with the large groove in the shaft during reassembly

18.11   Apply a thin coat of high-temperature grease to the shoe contact areas of the cam and pivot before installing the shoes

machine is frequently washed at car washes with high pressure nozzles). Before beginning, read through the entire Section to familiarize yourself with the procedure.

2 To do the job properly you will need some medium-weight, all purpose grease (do not use automotive wheel bearing grease), a large drift punch, a brass or plastic-tipped hammer, some clean solvent, a clean container and parts cleaning brush and a heat source such as a hot plate, propane torch or stove top burner. New seals and possibly new bearings will also be required.

Front wheel bearings

3 Begin by removing the wheel from the machine (refer to Section 9).
4 Remove the speedometer drive unit from the left side of the hub.
5 Remove the spacer, then using a large screwdriver pry the seal out of the right side of the hub. Cushion the hub with a shop towel to avoid damaging the area around the seal. Next, pry the seal out of the left side of the hub (photo) and remove the speedometer drive coupling washer. It is not absolutely necessary, but the disc should be removed from the hub as well.
6 Wipe all dirt and grease from the exposed portion of the bearings and the surrounding hub area.
7 The bearings are held in the hub by an interference fit and should never be driven out, as damage to both the hub and bearings will result. The bearings should be removed by carefully heating the hub, which

will cause it to expand and release the bearings.
8 To do this you will need a controllable heat source such as an electric hot plate or stove top burner. If a stove top burner is used, be sure to place a metal plate between the flame and the hub. A propane or welding torch can also be used, but extreme care must be taken to heat the hub evenly and keep the flame away from the bearing. If the bearing is overheated at all it will have to be replaced with a new one.
9 Set the wheel hub on the hot plate or over the burner and allow it to heat up. It is ready when a small drop of water placed on the hub next to the bearing boils away. Quickly remove the hub from the heat source. The bearings and spacer should fall out of place, so have a couple of shop towels laid out to cushion their fall. If not, give the hub a rap with a soft-faced hammer to dislodge them.
10 If a torch is used, remove one bearing at a time. Keep the torch moving to heat the hub evenly and do not apply any heat directly to the bearing. Never reuse a bearing that has been overheated.
11 Allow the bearings to cool, then wash them thoroughly with clean solvent to remove the old grease. Dry the bearings, then check them for roughness and binding, turning them slowly by hand. If any roughness is noted, they must be replaced with new ones. Most bearings will have some play between the races and the balls. If the play is excessive, replace them with new ones.
12 Pack the bearings with clean grease until they are approximately 2/3 full (photo). Since they are sealed bearings, the grease will have

18.13 Measuring the rear brake drum diameter with dial calipers

19.5 Prying the seal out of the front hub with a large screwdriver (note the towel used to protect the hub)

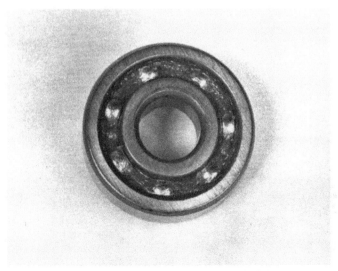

19.12 Pack the wheel bearings approximately 2/3 full of clean, high-quality grease (do not use automotive wheel bearing grease)

19.15 Installing the seal in the hub with a hammer and block of wood

to be forced in from the open side.

13 At this point, check to make sure that the bearing mounts in the hub are clean, then reheat the hub. Place the bearings in a refrigerator or freezer. This will cause the hub to expand and the bearings to contract, easing installation.

14 Position one of the bearings and push it in until it seats in the hub. Turn the wheel over and install the spacer, followed by the remaining bearing. Make sure the sealed sides of the bearings face out. Carefully tap around the entire circumference of the outer race of each bearing with the hammer and punch to ensure complete seating. *Do not hammer on the inner race, as the bearing will be damaged.*

15 After the hub has cooled, install the new seals. Apply a very thin coat of grease to the outer circumference of each seal, then tap them into place in the hub using a large socket (with the same diameter as the seal) or a block of wood and a hammer (photo). Apply pressure evenly to the outer edge of the seal. Be sure to lay the speedometer drive coupling washer in place next to the bearing before installing the seal in the left side of the hub.

16 Refer to Section 9 and install the wheel on the machine.

Rear wheel bearings

17 Refer to the appropriate Section and remove the rear wheel from the machine. Remove the bolts and separate the clutch hub from the wheel, then separate the large O-ring from the clutch hub.

18 Being careful not to inhale the hazardous dust, clean out the drum area with a stiff brush to remove the brake dust.

19 Refer to Paragraphs 6 through 15 for the actual bearing inspection and maintenance procedure. It's identical for both front and rear wheel bearings. Note that the left side bearings are mounted in the clutch hub.

20 After the bearings have been installed in the hubs, attach the clutch hub to the wheel and tighten the bolts to the specified torque. Follow a criss-cross pattern to avoid warping the hub and work up to the final torque in three steps.

21 Refer to the appropriate Section and install the wheel on the machine.

---

20   Tubeless tires — general note

1   Tubeless tires are used as standard equipment on US models. They are generally safer than tube-type tires but if problems do occur they require special repair techniques.

2   The force required to break the seal between the rim and the bead of the tire is substantial, and is usually beyond the capabilities of an individual working with normal tire irons.

3   Also, repair of the punctured tire and replacement on the wheel rim requires special tools, skills and experience that the average do-it-yourselfer lacks.

4   For these reasons, if a puncture or flat occurs with a tubeless tire, the wheel should be removed from the motorcycle and taken to a dealer service department or a motorcycle repair shop for repair or replacement of the tire.

Chapter 8 Electrical system

Contents

Specifications

Battery

Capacity.. 12 volt, 14 amp/hour
Specific gravity
    Fully charged................................ 1.280
    Minimum...................................... 1.260

Circuit fuse ratings

Main .... 30 amp
Headlight .... 10 amp
Signal lights .... 10 amp
Ignition system .... 10 amp
Electric cooling fan .... 10 amp

Starter motor

Armature coil insulation resistance .... 11.5 M ohms minimum
Brush length
    Standard..................................... 13/32 in/0.400 in (10 mm)
    Service limit .... 3/16 in/0.180 in (4.5 mm)
Brush spring pressure .... 31.7 ± 3.2 oz (900 ± 90 g)
Commutator diameter
    Standard..................................... 1.100 in (28 mm)
    Service limit .... 1.060 in (27 mm)
Mica undercut depth................................. 0.060 in (1.6 mm)

Alternator/voltage regulator

Alternator output .... 14 volts (20 amps) at 3000 rpm
No-load regulated voltage .... 14.5 ± 0.5 volts

Relay resistances

Starting circuit cut-off relay .... 100 ± 10 ohms
Electric fan relay.................................. 100 ± 10 ohms

Temperature gauge coil resistance .... 55 ± 2 ohms at 68°F

1 General information

The Yamaha Vision is equipped with a 12-volt electrical system. The components include a crankshaft mounted permanent magnet alternator and a solid state regulator/rectifier unit.

The regulator maintains the charging system output within the specified range to prevent overcharging and the rectifier converts the AC output of the alternator to DC current to power the lights and other components and to charge the battery.

The alternator consists of a multi-coil stator (bolted to the left-hand engine case) and a permanent magnet rotor. Mounted with the alternator are some of the components which comprise the TCI ignition system.

An electric starter mounted beneath the front of the engine is standard equipment. The starting system includes the motor, the battery, the solenoid, the starting circuit cut-off relay and the various wires and switches. If the engine STOP switch and the main key switch are both in the On position, the cut-off relay allows the starter motor to operate only if the transmission is in Neutral (Neutral switch on) or the clutch lever is pulled to the handlebar (clutch switch on) and the sidestand is up (sidestand switch on).

The majority of electrical system connections and some minor components are located behind the left side cover or coolant reservoir and inside the headlight shell.

Note: *Keep in mind that electrical parts, once purchased, cannot be returned. To avoid unnecessary expense, make very sure the faulty component has been positively identified before buying a replacement part.*

2 Electrical troubleshooting

A typical electrical circuit consists of an electrical component, the switches, relays, etc. related to that component and the wiring and connectors that hook the component to both the battery and the frame. To aid in locating a problem in any electrical circuit, complete wiring diagrams of each model are included at the end of this Chapter.

Before tackling any troublesome electrical circuit, first study the appropriate diagrams thoroughly to get a complete picture of what makes up that individual circuit. Trouble spots, for instance, can often be narrowed down by noting if other components related to that circuit are operating properly or not. If several components or circuits fail at one time, chances are the fault lies in the fuse or ground connection, as several circuits often are routed through the same fuse and ground connections.

Electrical problems often stem from simple causes, such as loose or corroded connections or a blown fuse. Prior to any electrical troubleshooting, always visually check the condition of the fuse, wires and connections in the problem circuit.

If testing instruments are going to be utilized, use the diagrams to plan where you will make the necessary connections in order to accurately pinpoint the trouble spot.

The basic tools needed for electrical troubleshooting include a test light or voltmeter, a continuity tester (which includes a bulb, battery and set of test leads) and a jumper wire, preferably with a circuit breaker incorporated, which can be used to bypass electrical components. Specific checks described later in this Chapter may also require an ammeter or ohmmeter.

Voltage checks should be performed if a circuit is not functioning properly. Connect one lead of a test light or voltmeter to either the negative battery terminal or a known good ground. Connect the other lead to a connector in the circuit being tested, preferably nearest to the battery or fuse. If the bulb lights, voltage is reaching that point, which means the part of the circuit between that connector and the battery is problem-free. Continue checking the remainder of the circuit in the same manner. When you reach a point where no voltage is present, the problem lies between there and the last good test point. Most of the time the problem is due to a loose connection. *Keep in mind that some circuits only receive voltage when the ignition key is in the On position.*

One method of finding short circuits is to remove the fuse and connect a test light or voltmeter in its place to the fuse terminals. There should be no load in the circuit. Move the wiring harness from side-to-side while watching the test light. If the bulb lights, there is a short to ground somewhere in that area, probably where insulation has rubbed

off a wire. The same test can be performed on other components in the circuit, including the switch.

A ground check should be done to see if a component is grounded properly. Disconnect the battery and connect one lead of a self-powered test light (such as a continuity tester) to a known good ground. Connect the other lead to the wire or ground connection being tested. If the bulb lights, the ground is good. If the bulb does not light, the ground is not good.

A continuity check is performed to see if a circuit, section of circuit or individual component is capable of passing electricity through it. Disconnect the battery and connect one lead of a self-powered test light (such as a continuity tester) to one end of the circuit being tested and the other lead to the other end of the circuit. If the bulb lights, there is continuity, which means the circuit is passing electricity through it properly. Switchs can be checked in the same way.

Remember that all electrical circuits are designed to conduct electricity from the battery, through the wires, switches, relays, etc. to the electrical component (light bulb, motor, etc.). From there it is directed to the frame (ground) where it is passed back to the battery. Electrical problems are basically an interruption in the flow of electricity from the battery or back to it.

3 Battery — inspection and maintenance

1   Most battery damage is caused by heat, vibration, and/or low electrolyte levels, so keep the battery securely mounted, check the electrolyte level frequently and make sure the charging system is functioning properly.

2   Refer to Chapter 1, *Tune-up and routine maintenance,* for electrolyte level and specific gravity checking procedures.

3   Check around the base inside of the battery for sediment, which is the result of sulfation caused by low electrolyte levels. These deposits will cause internal short circuits, which can quickly discharge the battery. Look for cracks in the case and replace the battery if either of these conditions is found.

4   Check the battery terminals and cable ends for tightness and corrosion. If corrosion is evident, remove the cables from the battery and clean the terminals and cable ends with a wire brush or knife and emery paper. Reconnect the cables and apply a thin coat of petroleum jelly to the connections to slow further corrosion.

5   The battery case should be kept clean to prevent current leakage, which can discharge the battery over a period of time (especially when it sits unused). Wash the outside of the case with a solution of baking soda and water. *Do not get any baking soda solution in the battery cells.* Rinse the battery thoroughly, then dry it.

6   If acid has been spilled on the frame or battery box, neutralize it with the baking soda and water solution, dry it thoroughly, then touch up any damaged paint. Make sure the battery vent tube is directed away from the frame and is not kinked or pinched.

7   If the motorcycle sits unused for long periods of time, disconnect the cables from the battery terminals. Refer to Section 4 and charge the battery approximately once every month.

4 Battery — charging

1   If the machine sits idle for extended periods or if the charging system malfunctions, the battery can be charged from an external source.

2   To properly charge the battery, you will need a charger of the correct rating, an hydrometer, a clean rag and a syringe for adding distilled water to the battery cells.

3   The maximum charging rate for any battery is 1/10 of the rated amp/hour capacity. As an example, the maximum charging rate for the 14 amp/hour battery would be 1.4 amps. If the battery is charged at a higher rate, it could be damaged.

4   Do not allow the battery to be subjected to a so-called quick charge (high rate of charge over a short period of time) unless you are prepared to buy a new battery.

5   When charging the battery, always remove it from the machine and be sure to check the electrolyte level before hooking up the charger. Add distilled water to any cells that are low.

6   Loosen the cell caps, hook up the battery charger leads (red to positive, black to negative), cover the top of the battery with a clean

rag, then, and only then, plug in the battery charger. **Warning:** *Remember, the gas escaping from a charging battery is explosive, so keep open flames and sparks well away from the area. Also, the electrolyte is extremely corrosive and will damage anything it comes in contact with.*

7   Allow the battery to charge until the specific gravity is as specified (refer to Chapter 1 for specific gravity checking procedures). *The charger must be unplugged and disconnected from the battery when making specific gravity checks.* If the battery overheats or gases excessively, the charging rate is too high. Either disconnect the charger or lower the charging rate to prevent damage to the battery.

8   If one or more of the cells do not show an increase in specific gravity after a long slow charge, or if the battery as a whole does not seem to want to take a charge, it is time for a new battery.

9   When the battery is fully charged, unplug the charger first, then disconnect the leads from the battery. Install the cell caps and wipe any electrolyte off the outside of the battery case.

5 Fuses — inspection and replacement

1   The fuse block is located under the seat (photo). The fuse block is protected by a plastic cover, which snaps into place. It contains fuses (and spares) which protect the main, headlight, signal light and ignition system wiring and components from damage caused by short circuits.

2   If a fuse blows, be sure to check the wiring harnesses very carefully for evidence of a short circuit. Look for bare wires and chafed, melted or burned insulation. If a fuse is replaced before the cause is located, the new fuse will blow immediately.

3   Never, under any circumstances, use a higher rated fuse or bridge the fuse block terminals, as damage to the electrical system could result.

4   Occasionally a fuse will blow or cause an open circuit for no obvious reason. Corrosion of the fuse ends and fuse block terminals may occur and cause poor fuse contact. If this happens, remove the corrosion with a wire brush or emery paper, then spray the fuse end and terminals with electrical contact cleaner.

6 Lighting system — check

1   The battery provides power for operation of the headlight, taillight, brake light, license plate light and instrument cluster lights. If none of the lights operate, always check battery voltage before proceeding. Low battery voltage indicates either a faulty battery, low battery electrolyte level or a defective charging system. Refer to Chapter 1 for battery checks and Section 28 for charging system tests. Also, check the fuse condition and replace any blown fuses with new ones.

Headlight

2   If the headlight is out when the engine is running, check the bulb first (use jumper wires to connect it directly to the battery terminals), then check for battery voltage (12 volts) at both sides of the headlight fuse (photo) with the key On. If no voltage is indicated, check the fuse and the main (key) switch wiring harness connector (inside the headlight shell). Twelve (12) volts should be indicated at the brown wire terminal coming out of the main switch. If no voltage is present at the brown wire terminal, check the main switch and the wiring from the battery to the switch.

3   If voltage is present at the headlight fuse, remove the right handlebar switch (it is held in place with two screws) and check for battery voltage at the red/yellow wire where it is connected to the starter switch (photo) with the key On. You will have to remove the two small screws and separate the wire guide and shield from the switch. If no voltage is indicated, check the wiring between the starter switch and the headlight fuse for an open circuit. **Note:** *Make sure the headlight fuse is in good condition and check for voltage at both fuse terminals.*

4   If voltage is present at the red/yellow wire, check for voltage at the blue/black wire of the dimmer switch (this will require removal of the left handlebar switch, which is held in place with two screws). If no voltage is indicated, check the starter switch and the wire between the starter switch and dimmer switch.

5   If voltage is present at the blue/black wire, check for battery voltage at the green wire and the yellow wire. If no voltage is indicated at either wire, the dimmer switch is defective and must be replaced with a new one.

5.1   Fuse block location

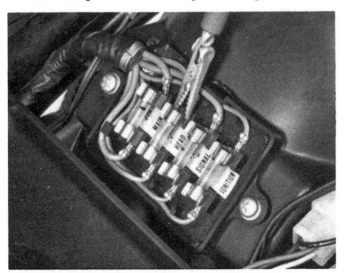

6.2   Checking for battery voltage at the headlight fuse terminal (the other voltmeter lead is attached to the negative terminal of the battery)

6.3   Check for battery voltage by touching the positive (red) voltmeter lead to the red/yellow wire solder joint (arrow) of the starter switch

6   If voltage is indicated at the green and yellow wires (depending on the position of the dimmer switch), check for voltage at the green and yellow wire terminals in the headlight bulb connector. If no voltage is indicated, there is an open circuit or a poor connection between the dimmer switch and the headlight.

7   If voltage is present at the headlight connector, check the connector itself and the headlight ground connection for corrosion (use a jumper wire to create a good ground circuit).

Taillight/license plate light

8   If the taillight fails to work, check the bulb and the bulb terminals first, then check for battery voltage at the blue wire in the taillight. If voltage is present, check the ground circuit for an open or poor connection.

9   If no voltage is indicated, check the wiring between the taillight and the main (key) switch, then check the switch.

Brake light

10   If the brake light does not work when the pedal or lever is depressed, refer to Chapter 1 and make sure the switch is properly adjusted, then check the signal light fuse and the bulbs (see Section 10 for bulb removal procedures). If the bulb(s) appear to be in good condition, remove them from the holders and check for battery voltage at the yellow wire terminal(s). The main (key) switch must be On and the brake lever or pedal must be depressed. If voltage is present, check the ground circuit for an open or poor connection.

11   If no voltage is indicated, locate the brake light switch wiring connectors and check for battery voltage at the brown wire connections. If no voltage is present, check the main fuse, the main switch and the wires between the main switch and the brake light switches.

12   If battery voltage is indicated, then the brake light switch is defective or the wiring between the switch and the taillight has an open or poor connection.

Neutral indicator light

13   If the neutral light fails to operate when the transmission is in Neutral, check the signal light fuse and the bulb (see Section 15 for bulb removal procedures). If the bulb and fuse are in good condition, check for battery voltage at the blue wire attached to the neutral switch on the left side of the engine. If battery voltage is present, refer to Section 20 for the neutral switch check and replacement procedures.

14   If no voltage is indicated, check the wiring between the signal light fuse and the neutral light and between the neutral light and the neutral switch for open circuits and poor connections.

Oil pressure warning light

15   If the oil pressure warning light fails to operate properly, check the oil level and make sure it is correct.

16   If the oil level is correct, disconnect the black/red wire from the oil pressure sending unit on the lower left side of the engine. Turn the main switch On and ground the end of the wire on the engine case. If the light comes on, the oil pressure sending unit is defective and must be replaced with a new one (only after draining the engine oil).

17   If the light does not come on, check the wiring between the oil pressure sending unit and the light and between the light and the signal light fuse.

---

7   Headlight bulb – replacement

1   The yamaha Vision is equipped with a sealed beam headlight. When replacement is necessary, the reflector, glass and bulb are replaced as an entire unit. **Note:** *RK and European models have replaceable bulbs.*

2   On RK models, remove the plugs on the fairing covering the headlight rim mounting screws. On all models, remove the two headlight rim assembly mounting screws and spacers from the headlight shell (photo).

3   Carefully pull out on the bottom of the headlight rim, then pull down to release the rim and beam assembly from the headlight shell. Unplug the electrical connector.

RJ models only

4   Disengage the spring clip (photo). Remove the two sealed beam adjusting screws (photo) then lift the sealed beam out of the rim.

5   Lay the new beam in place and install the adjusting screws and the spring clip.

All others

6   Turn the bulb holder counterclockwise and remove the defective bulb. Slip a new bulb into position and secure it with the bulb holder. Caution: Do not touch the glass part of the bulb with your fingers – it must be kept oil-free.

All models

7   Plug in the electrical connector. Slip the tab on the upper part of the rim into the recess in the headlight shell and push in carefully on the bottom of the rim assembly until the screw holes line up with the holes in the headlight shell. If resistance is felt, move the wiring components inside the headlight shell to the side to allow room in the center of the shell for the connector.

8   Install the spacers and mounting screws and tighten the screws securely.

9   Adjust the headlight aim by referring to Chapter 1.

---

8   Headlight assembly – removal and installation

1   Complete removal of the headlight assembly requires removal of the headlight and disconnection of the wiring inside the headlight shell.

2   Refer to Section 7 and remove the headlight rim and beam assembly. Remove the upper fairing if necessary.

3   Unplug all the wiring connectors inside the headlight shell. The wiring connectors and wires are color-coded to ease reassembly.

4   Remove the two mounting bolts/nuts, lift off the headlight shell

6.4   Battery voltage should be indicated at the blue/black wire solder joint (arrow) of the dimmer switch

7.2   Removing the headlight rim assembly mounting screws (RJ model)

and thread the wiring harnesses through the openings in the shell.
5   Installation is the reverse of removal.

9   Headlight aim — check and adjustment

Refer to Chapter 1, *Tune-up and routine maintenance,* for headlight aim checking and adjustment procedures.

10   Taillight and turn signal bulbs — replacement

1   Bulb replacement for the turn signals is very simple and requires only the removal of the plastic lens, which is held in place with screws (photo), and the bulb.
2   To remove the bulb, push in and turn it counterclockwise (photo). Check the socket terminals for corrosion and clean them if necessary. Line up the pins on the new bulb with the slots in the socket, push in and turn the bulb clockwise until it locks in place. It is a good idea to use a paper towel or a dry cloth when handling the new bulb to prevent injury if the bulb should break and to increase bulb life.
3   When replacing the lens, make sure the rubber gasket is not kinked or pinched and do not overtighten the screws.
4   To remove the taillight bulbs, raise the seat and slide the tool kit

7.4a   Disengaging the sealed beam spring clip

7.4b   Removing the sealed beam adjusting screws

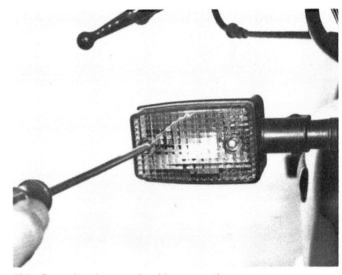

10.1   Removing the turn signal lens mounting screws

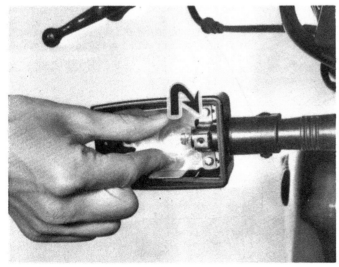

10.2   Removing the turn signal bulb

10.5   Separating the bulb holder from the taillight assembly

storage compartment out of the tail assembly.

5   Turn the bulb holders counterclockwise (photo) until they stop, then pull straight out to remove them from the taillight housing. The bulbs can be removed from the holders by turning them counterclockwise and pulling straight out.

6   Check the socket terminals for corrosion and clean them if necessary. Line up the pins on the new bulb with the slots in the socket, push in and turn the bulb clockwise until it locks in place. **Note:** *The pins on the bulb are offset so it can only be installed one way.* It is a good idea to use a paper towel or dry cloth when handling the new bulb to prevent injury if the bulb should break and to increase bulb life.

7   Make sure the rubber gaskets are in place and in good condition, then line up the tabs on the holder with the slots in the housing and push the holder into the mounting hole. Turn it clockwise until it stops to lock it in place. **Note:** *The tabs and slots are two different sizes so the holders can only be installed one way.*

8   Replace the tool kit storage compartment and lower the seat.

---

11   Turn signal assemblies — removal and installation

1   The turn signal assemblies can be removed individually in the event of damage or failure.

2   To remove the front turn signal assemblies, remove the headlight beam and rim assembly (refer to Section 7) and unplug the turn signal wiring lead bullet connectors (two for each signal). Unscrew the turn signal assembly mounting nuts and pull them out of the headlight bracket.

3   The turn signal lights are clamped to the stalks and can be removed separately by taking out the screws and sliding the lights off the stalks. This is best done with the light assemblies in place on the machine.

4   When installing the turn signals, be sure to hook up the wiring leads correctly. They are color-coded to ease reassembly.

5   The rear turn signal assemblies can be removed by lifting the seat, disconnecting the wiring lead bullet connectors and removing the nuts on the inside of the stalks.

6   Again, the lights are clamped to the stalks and can be removed separately with the assemblies in place on the machine.

---

12   Turn signal system — check

1   The battery provides power for operation of the signal lights, so if they do not operate, always check the battery voltage and specific gravity first. Low battery voltage indicates either a faulty battery, low electrolyte level or a defective charging system. Refer to Chapter 1 for battery checks and Sections 27 and 28 for charging system tests. Also, check the signal light fuse and replace it with a new one if it is blown.

Right side lights do not flash

2   If the turn signal lights fail to operate, check the condition of the bulb(s) first (see Section 10 for bulb removal procedures).

3   If the bulb(s) are in good condition, check for battery voltage at the dark green wire in the light socket (the main switch must be On). If battery voltage is indicated, check the ground circuit for an open or poor connection. If no battery voltage is indicated, check the wiring between the switch and the light(s), then check the switch.

Left side lights do not flash

4   The procedure is the same as for the right side lights but note that the 'hot' lead at the socket is dark brown rather than dark green.

Both right and left signal lights do not flash

5   Remove the left handlebar switch (it is held in place with two screws) and check for battery voltage at the brown/white wire solder connection at the signal light switch (photo) (you will have to remove the wire guide and plastic shield to gain access to the wire connection). If voltage is indicated, check the switch.

6   If no voltage is indicated, check for battery voltage at the brown wire terminal of the turn signal relay (located behind the TCI unit on the right side of the motorcycle). If voltage is indicated, check the wire (brown/white) between the relay and the switch. If the wire is in good condition, the relay is defective and must be replaced with a new one. If no voltage is indicated, check the wire (brown) between the signal light fuse and the relay. Also, make sure that voltage is present at the signal light fuse terminals.

12.5   Check for battery voltage by touching the positive (red) voltmeter lead to the brown/white wire solder joint (arrow) of the signal light switch

Signal lights fail to cancel

7   The self-cancelling unit turns off the signal lights after either ten seconds of operation or after the motorcycle has travelled 490 feet (150 meters), whichever comes first. The light may also be cancelled manually by pushing in on the switch lever. **Note:** *If the signal lights flash when the switch is in the Left or Right turn position, but stop flashing if the switch is moved to Off, the self-cancelling unit is defective and should be replaced with a new one.*

8   If the signal lights fail to cancel, disconnect the wiring harness from the cancelling unit (located on the backside of the TCI unit bracket) and ground the yellow/green wire with a jumper lead. If the signal lights now operate normally (the main switch must be On), the flasher relay, the bulbs, the handlebar switch and the circuit wires are all in good condition. However, the cancelling unit, handlebar switch reset circuit and the speedometer sensor circuit may be faulty. To check further, proceed as follows.

9   Connect an ohmmeter to the white/green and black wire terminals in the wiring harness side of the cancelling unit connector. Place the ohmmeter selector switch in the Rx100 range. Spin the front wheel (have an assistant sit on the rear of the seat to raise the wheel off the ground) and watch the ohmmeter needle. It should swing back and forth between zero and infinite ohms. If it does, the speedometer sensor circuit is in good condition. If it doesn't, the sensor wiring (white/green and black wires) may be open.

10   Move the ohmmeter leads to the yellow/red wire terminal in the harness connector and a good ground, then operate the signal light switch. When the switch is in the Left or Right turn positions, the ohmmeter should indicate zero (0) ohms. When the switch is in the Off position, the ohmmeter should indicate infinite ohms. If it doesn't, check the handlebar switch and the wiring harness.

11   If no defects have been found to this point, the self-cancelling unit is defective and should be replaced with a new one.

---

13   Brake light switches — check and adjustment

Refer to Chapter 1 for the procedure to follow when checking and adjusting the brake light switches.

---

14   Instrument cluster — removal and installation

1   The instrument cluster is attached to the upper triple clamp at two points. **Note:** *Although it is not absolutely necessary, it may be a good idea to remove the headlight assembly mounting bolts so the headlight shell can be pushed down to provide room to work on the instrument cluster.*

2   Unscrew the fitting and separate the speedometer cable from the rear of the speedometer.

15.1   Headlight bracket mounting bolt locations

15.2a   Instrument cluster bottom cover mounting screws

3   Follow the instrument cluster wiring harness into the headlight shell and unplug the two connectors, then push the harness out through the opening in the back of the headlight.
4   Remove the two mounting nuts and separate the instrument cluster from the upper triple clamp.
5   Installation is basically the reverse of removal. Be sure to install the rubber grommets and the large washers in their correct locations.

15   Instrument and warning light bulbs — replacement

1   To replace the instrument light bulbs, separate the plastic horn cover from the lower triple clamp (it is held in place with two bolts) then remove the horn and the two headlight bracket mounting bolts (photo). Pull down on the headlight assembly to provide access to the instrument cluster bottom cover screws.
2   Remove the three screws (photo) and separate the bottom cover from the instrument cluster. Pull the appropriate rubber bulb socket out of the instrument cluster panel (photo), then pull the bulb out of the socket. If the socket contacts are dirty or corroded, they sould be scraped clean and sprayed with electrical contact cleaner before new bulbs are installed. Carefully push the new bulb into position, then install the socket. When installing the instrument cluster panel, do not overtighten the screws (they thread into plastic).
3   Push up on the headlight assembly and engage the rubber mounts in the holes in the upper triple clamp, then install the bolts and tighten them securely. Attach the horn cover to the triple clamp and tighten the bolts, then install the plastic covers over the bolts.
4   To replace the warning light bulbs, depress the tabs on the under side of the warning light panel and separate the panel from the upper triple clamp by pushing it up. Turn the handlebars all the way to the right to provide room to work.
5   Pull the appropriate rubber bulb socket out of the bottom of the panel, then pull the bulb out of the socket. If the socket contacts are dirty or corroded, they should be scraped clean and sprayed with electrical contact cleaner before new bulbs are installed. Carefully push the new bulb into position, then install the socket in the panel. Position the panel in the triple clamp opening and push it straight down to lock it in place.

16   Ignition main (key) switch — check

1   Although it is not common, switch malfunctions occasionally do occur. Proper operation of the ignition switch can be verified with a continuity test light or an ohmmeter. If the switch is found to be defective, it must be replaced with a new one, as repair is not possible.
2   Remove the headlight rim and beam assembly (refer to Section 7). Trace the wiring harness back from the ignition switch and unplug the block connector inside the headlight shell.

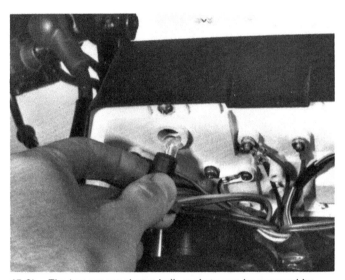

15.2b   The instrument cluster bulb sockets can be removed by gently pulling on the rubber base

Switch position	Wire color		
	R	Br	L/Y
ON	●	●	●
OFF			
LOCK			
P (parking)	●		●

Fig. 8.1   Ignition main (key) switch terminal continuity (Sec 16)

3   Using the ohmmeter or test light, check for continuity between the terminals of the ignition switch harness with the key in the various switch positions (refer to Figure 8.1). Continuity should exist between the terminals connected by a solid line when the switch is in the indicated position.

17 Ignition main (key) switch — removal and installation

1   Refer to Section 7 and remove the headlight rim and beam assembly. Trace the ignition switch harness into the headlight shell, unplug the block connector and push it out through the back of the shell.
2   Turn the handlebars all the way to the left, then remove the two bolts that attach the switch to the upper triple clamp. Pull down and back on the switch to separate it from the mount.
3   Installation is the reverse of removal. Be sure to check the new switch for proper operation before installing it.

18 Handlebar switches — check

1   Generally speaking, the switches are reliable and trouble-free. Most troubles, when they do occur, are caused by dirty or corroded contacts, but wear and breakage of internal parts is a possibility that should not be overlooked. If breakage does occur, the entire switch and related wiring harness will have to be replaced with a new one, since individual parts are not usually available.
2   The switches can be checked for continuity with an ohmmeter or a continuity test light. Always disconnect the battery ground cable, which will prevent the possibility of a short circuit, before making the checks.
3   Refer to Section 7 and remove the headlight rim and beam assembly. Trace the wiring harness of the switch in question into the headlight shell and unplug the harness connectors.
4   Using the ohmmeter or test light, check for continuity between the terminals of the switch harness with the switch in the various positions (refer to the appropriate Figure). Continuity should exist between the terminals connected by a solid line when the switch is in the indicated position.
5   If the continuity check indicates that a problem exists, refer to Section 19, disassemble the switch and spray the switch contacts with electrical contact cleaner. If they are accessible, the contacts can be scraped clean with a knife or polished with crocus cloth. If switch components are damaged or broken, it will be obvious when the switch is disassembled.

19 Handlebar switches — removal and installation

1   The handlebar switches are composed of two halves that clamp around the bars. They are easily removed for cleaning or inspection by taking out the clamp screws and pulling the switch halves away from the handlebars.
2   To completely remove the switches, the wiring harness should be unplugged and pulled out of the headlight shell. The right side switch must be separated from the throttle cable also.
3   When installing the switches, make sure the wiring harnesses are properly routed to avoid pinching or stretching the wires.

20 Neutral switch — check and replacement

1   The neutral switch, which is threaded into the crankcase on the lower left side of the engine (directly behind the sidestand bracket), operates an indicator light in the panel when the transmission is in Neutral. More importantly, it is connected with the starter solenoid and will allow the engine to be started only if the transmission is in Neutral,

Switch position	Wire color	
	R/W	R/W
RUN	●———————●	
OFF		

Fig. 8.2   Engine stop switch terminal continuity (Sec 18)

Switch position	Wire color	
	L/W	Ground
PUSH	●———————●	
OFF		

Fig. 8.3   Electric starter switch terminal continuity (Sec 18)

Switch position	Wire color				
	Ch	Br/W	Dg	Y/R	Ground
L	●——●			●——●	
L → N	●——●				
N → Push					
R → N		●——●			
R		●——●		●——●	

Fig. 8.4   Turn signal switch terminal continuity (Sec 18)

Switch position	Wire color		
	Y	L/Y	G
HI	●——●		
LO		●——●	

Fig. 8.5   Headlight dimmer switch terminal continuity (Sec 18)

Switch position	Wire color	
	P	Ground
PUSH	●———————●	
OFF		

Fig. 8.6   Horn switch terminal continuity (Sec 18)

unless the clutch is disengaged.

2   It can be checked with an ohmmeter or a continuity test light. Disconnect the switch wire, then touch one ohmmeter or light probe to the switch and touch the other probe to the engine crankcase. The meter or light should indicate continuity when the transmission is in Neutral and no continuity when the transmission is in gear.

3   If the switch is defective, it must be replaced with a new one. You may have to remove the footpeg and sidestand bracket to gain access to the switch. **Note:** *Before unscrewing the switch from the crankcase, tip the motorcycle slightly to the right and support it so it cannot fall. this will prevent oil from running out of the hole in the crankcase when the switch is removed.* Install and tighten the new one after coating the threads with thread sealant *(do not use teflon tape)*, then hook up the switch wire.

21   Horn — adjustment

1   The horn, which is attached to a bracket on the lower triple clamp, can be adjusted to produce the best sound.

2   Loosen the adjusting screw locknut, which is located on the back side of the horn, then turn the key to the On position.

3   Push the horn button and turn the adjusting screw in or out until the desired sound is obtained. Retighten the locknut.

22   Starting circuit cut-off relay — check

1   The starting circuit cut-off relay is located behind the TCI unit mounting bracket. To gain access to it, lift the seat and remove the right side cover. The coolant reservoir and TCI unit can be moved aside after removing the screw that holds the top of the reservoir in place (you may have to disconnect the upper reservoir hose, but be sure to leave the lower hose in place).

2   The relay is attached to a frame tab by a rubber cushion mount (photo). It can be identified by examining the wiring at the connector (look for a connector with a black/yellow, light blue and two red/white wires attached to it).

3   Pull the relay and rubber mount off the frame tab, then depress the connector lock and separate the connector from the relay.

4   Using an ohmmeter, check the resistance of the relay coil windings. Hook up the ohmmeter leads as indicated in Photo 22.4. The resistance should be 100 ohms; if it is not, replace the relay with a new one. If the resistance is as specified, proceed to Step 5.

5   Connect the motorcycle's battery and an ohmmeter to the relay terminals as shown in Photo 22.5a (use jumper leads with automotive-type female tab connectors (photo) to hook up the battery). With the battery *connected*, the ohmmeter should read zero (0) ohms. With the battery *disconnected*, the ohmmeter should read infinite resistance.

22.2   Starting circuit cut-off relay location (arrow)

22.4   Check the coil winding resistance by attaching the ohmmeter leads to the terminals indicated (arrows)

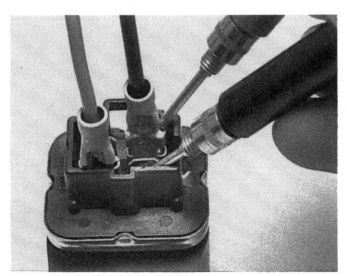

22.5a   Checking the relay operation by applying 12 volts (tab connectors) and checking for continuity between the right hand terminals

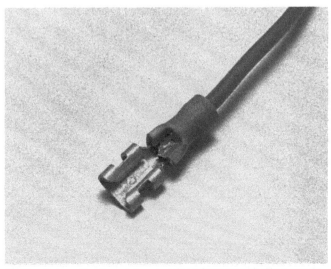

22.5b   Automotive-type tab connectors (female) will allow you to apply voltage to the relay terminals without accidentally causing a short circuit

Note: *A continuity test light can be used in place of an ohmmeter. If a light is used, it should glow when the battery is connected and go out when it is disconnected.*

6   If the relay fails the check, replace it with a new one.

7   To check the relay diode, connect the ohmmeter leads as shown in the accompanying Figure (red lead to the upper left terminal, black lead to the lower left terminal). Place the selector switch in the Rx1 position. the ohmmeter will indicate very low resistance (if you are using a genuine Yamaha pocket tester, the reading will be 9.5 ohms; other ohmmeters will give different readings but they will all be relatively low). Reverse the ohmmeter leads as shown in the Figure (red lead to lower left terminal, black lead to upper left terminal). the ohmmeter should now indicate very high or infinite resistance. If the relay fails the test, replace it with a new one.

23   Starter solenoid — check

1   Remove the left side cover and locate the solenoid. Turn the main (key) switch to On, the engine stop switch to Run and make sure the transmission is in Neutral. Push the start button and listen for a 'click' from the solenoid. If no sound is heard, turn the switch off and check the connector in the wiring harness attached to the solenoid (red/white and blue/white wires). Check for battery voltage at the wire terminals (the switch must be On). If no voltage is indicated, check the starting circuit cut-off relay, the stop switch, the ignition fuse, the main switch,

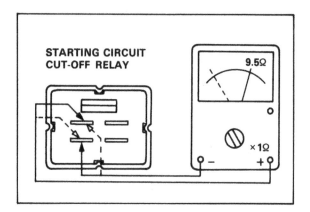

Fig. 8.7   Ohmmeter lead connections for checking the starting circuit cut-off relay diode (Sec 22)

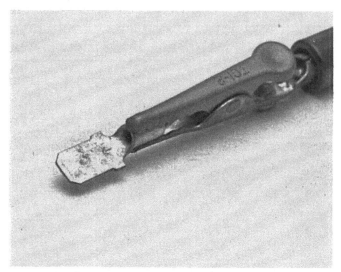

23.4b   The ohmmeter leads can be attached to the tab connector, then the connector can be inserted into the terminal

the main fuse and all wiring and connections, in that order.

2   If voltage is indicated at the red/white wire terminal and not the blue/white terminal, check the solenoid coil as follows.

3   Make sure the main switch is Off, then disconnect the solenoid wiring harness connector. Attach the leads from an ohmmeter to the red/white and blue/white wires leading to the solenoid (photo) and place the ohmmeter selector switch in the Rx1 position. Use automotive-type male tab connectors (photo) to attach the ohmmeter leads to the wire terminals. The ohmmeter should indicate 3.5 ohms. If the reading is greater than 3.5 ohms, or if it is zero (0), the solenoid is defective and should be replaced with a new one.

24   Starter solenoid — removal and installation

1   Make sure the main switch is Off, then remove the left side cover and disconnect the negative cable from the battery. Remove the plastic battery cover.

2   Carefully pry up the plastic covers on the solenoid and remove both large cables (photo).

3   Unplug the small wire connector, then slide the solenoid off the mounting tabs on the battery holder.

4   Installation is the reverse of removal. Reconnect the negative battery cable after all the other electrical connections are made.

23.4a   Checking the starter solenoid coil resistance (note the use of male automotive-type tab connectors to tap into the connector terminals)

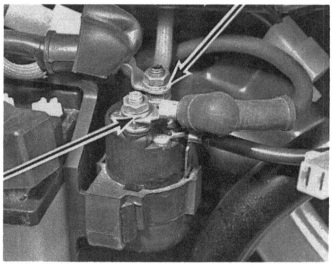

24.2   Detach the large cables (arrows) from the solenoid before removing it from the mount

25 Starter motor — removal and installation

1 Place the motorcycle on the centerstand and make sure the main switch is Off.
2 Remove the left side cover and disconnect the negative battery cable from the battery.
3 Pull back the rubber boot and disconnect the battery lead from the starter motor terminal (photo).
4 Remove the two mounting bolts (photo) and pull the starter straight out of the engine case.
5 Installation is the reverse of removal. Clean the mounting area on the engine case and apply a thin coat of oil or grease to the O-ring on the starter before sliding it back into place.

26 Starter motor — disassembly, inspection and reassembly

Disassembly

1 Remove the two long case screws (photo) and separate the end covers from the motor. Don't lose the thrust washers that fit over the ends of the armature. The armature is free to slip out of the case, so support it securely.
2 Remove the brush lead attaching bolt nut (photo) and carefully lift

25.3 Removing the battery lead from the starter motor terminal

25.4 Removing the upper starter motor mounting bolt

26.1 Removing the screws that hold the starter motor components together

26.2 Removing the nut from the brush lead terminal bolt

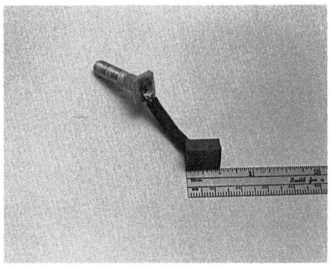
26.6 Measuring the brush length with a ruler

26.7a   Check the commutator segments (arrow) for excessive wear, damage and discoloration

the brush holder assembly off.

3   Withdraw the armature and end plate from the motor case, then carefully separate the armature and plate. Note how the thrust washers and spacer are installed.

4   Pull back the spring and slide the brush attached to the bolt out of the holder.

5   Clean the motor components thoroughly to remove brush residue, oil, and dirt.

Inspection

6   The parts of the starter motor that most likely will require attention are the brushes. Measure the length of the one attached to the bolt (photo) and compare the results to the Specifications. If either one or both brushes are worn beyond the specified limits, replace the brush holder assembly with a new one. If the brushes are not worn excessively, cracked, chipped, or otherwise damaged, the brush holder assembly may be reused. Reinstall the brush in the holder, then slip the bolt through the end cover and attach the nut.

7   Inspect the commutator for scoring, scratches and discoloration. The commutator can be cleaned and polished with crocus cloth, but do not use sandpaper or emery paper. After cleaning, wipe away any residue with a cloth soaked in an electrical system cleaner or denatured alcohol. Measure the commutator diameter and compare it to the Specifications (photos). If it is less than the service limit, the motor

26.7b   Measuring the commutator diameter with a dial caliper

26.9a   Checking for continuity between the commutator bars

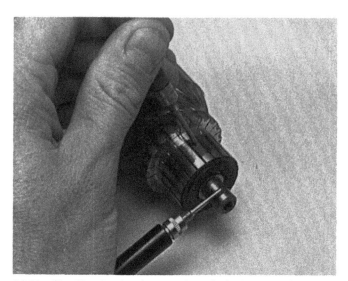

26.9b   Checking for the absence of continuity between the commutator bars and the armature shaft

26.10   Checking for the absence of continuity between the brush lead terminal bolt and the end cover

must be replaced with a new one.

8   Yamaha does supply specifications for under cutting the mica insulators between commutator segments, which implies that the commutator can be turned if the armature is in otherwise good condition. If the armature is shorted internally or has an open winding, the entire motor must be replaced with a new one (armatures are not available separately). This is a rather costly solution, so it may be very worthwhile to check with an automotive electrician concerning the possibility of reconditioning the commutator so the armature can be reused.

9   Using an ohmmeter or a continuity test light, check for continuity between the commutator bars (photo). Continuity should exist between each bar and all of the others. Also, check for continuity between the commutator bars and the armature shaft (photo). There should be no continuity between the commutator and the shaft. If the checks indicate otherwise, the armature is defective.

10   Check for continuity between the motor end cover and the battery cable terminal bolt (photo). There should be no continuity. Continuity should exist between the battery cable terminal and the brush lead. If the checks indicate otherwise, the insulator between the bolt and cover is defective.

11   Check the starter drive gear for worn, cracked, chipped and broken teeth. If the gear is damaged or worn, it can be separated from the shaft after removing the snap-ring (photo). When reinstalling the snap-ring, make sure the sharp edge faces out and do not expand it any more

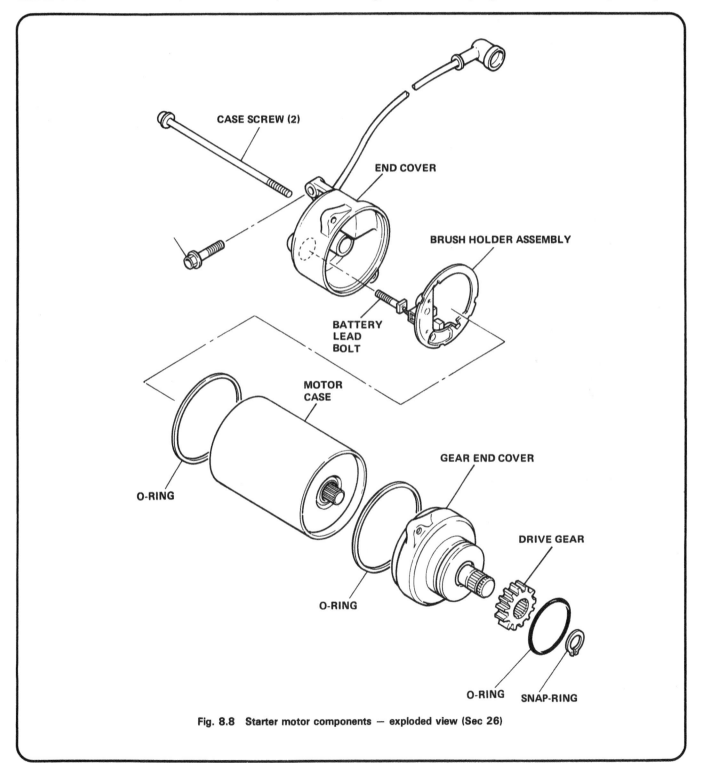

Fig. 8.8   Starter motor components — exploded view (Sec 26)

than necessary to slip it onto the shaft. Make sure it is seated in the groove.

12 Check the reduction gears inside the cover (photo) for wear and damage. Make sure the output shaft rotates smoothly without binding or excessive side play. If the gears and/or bearing require replacement, a new motor will be required.

13 Check the armature shaft ends and the bushings in the cover and end plate (photo) for excessive wear and scoring. If wear or damage has occurred, a new motor will be required.

Reassembly

14 Slip the thrust washers and spacer onto the gear end of the armature in the correct order (photo), then apply a thin coat of high-temperature grease to the bushing and install the end plate in the motor case (photo).

15 Slip the armature into the motor and through the bushing in the end plate (make sure the thrust washers do not fall off the shaft as the armature passes through the motor — they are attracted by the magnets in the case).

16 Slip the thrust washers and spacer over the commutator end of the armature in the correct order (photo).

17 Apply a thin coat of high-temperature grease to the bushing in the end cover, make sure the brush holder tab is properly seated in the groove, then attach the cover to the motor by slipping the brushes carefully over the commutator and engaging the armature end in the

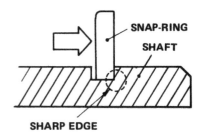

Fig. 8.9   Install the snap-ring on the end of the motor shaft with the sharp edge facing the end of the shaft (Sec 26)

cover bushing. The small tab on the brush holder must fit into the recess in the motor case (photo) (the recess is adjacent to the raised alignment mark on the outside of the case).

18 Lubricate the reduction gear backing plate and the gear on the end of the armature shaft, then attach the cover to the motor. The dimple on the case must be aligned with the groove in the reduction ring gear as the cover is installed (photo).

19 Make sure the covers are seated properly on the motor case with the marks aligned, then install and tighten the two long screws (photo).

20 Check the large O-ring on the gear end of the motor. If it is damaged or cracked, replace it with a new one before installing the motor.

---

27   Charging system testing — general note

1   If the performance of the charging system is suspect, the system as a whole should be checked first, followed by testing of the individual components (the alternator, the rectifier and the regulator). **Note:** *Before beginning the checks, make sure that the battery is fully charged and that all system connections are clean and tight.*

2   Checking the output of the charging system and the performance of the various components within the charging system requires the use of special electrical test equipment. A voltmeter and ammeter or a multimeter are the absolute minimum tools required. In addition, an ohmmeter is generally required for checking the remainder of the system.

3   When making the checks, follow the procedures carefully to prevent incorrect connections or short circuits, as irreparable damage to electrical system components may result if short circuits occur. Because of the special tools and expertise required, it is recommended that the job of checking the charging system be left to a dealer service department or a reputable motorcycle repair shop.

26.11   Removing the snap-ring in order to slide off the gear

26.12   Check the reduction gears and shafts for wear and damage

26.13   Check the bushings (arrows) for wear and damage

26.14a   Install the thrust washers and spacer with the spacer in the middle

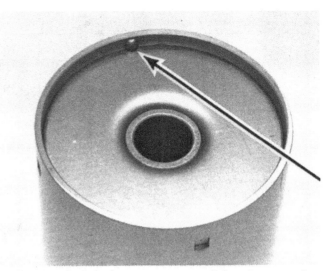

26.14b   The dimple in the motor case must fit into the cut-out in the end plate (arrow)

26.16   Install the thrust washers and spacer with the spacer in the middle

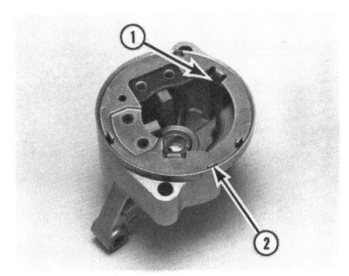

26.17   Make sure the large brush holder tab is positioned in the groove (1) and align the small tab (2) with the recess in the case

26.18   The dimple on the case must be aligned with the groove in the ring gear (arrows) as the cover is attached to the motor

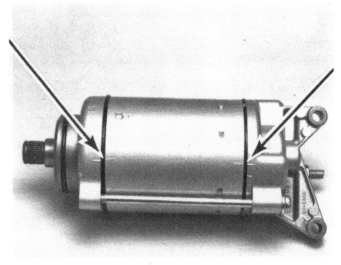

26.19   Make sure the index marks are aligned before tightening the screws

28 Charging system — output test

Caution: *Never disconnect the battery cables from the battery while the engine is running. If the battery is disconnected, the alternator and regulator/rectifier will be damaged.*

1   To check the charging system output, you will need a voltmeter or a multimeter with a voltmeter function.

2   The battery must be fully charged (charge it from an external source if necessary) and the engine must be at normal operating temperature to obtain an accurate reading.

3   Attach the positive (red) voltmeter lead to the positive (+) battery terminal and the negative (black) lead to the battery negative (-) terminal. The voltmeter selector switch (if so equipped) must be in a DC volt range greater than 15 volts.

4   Start the engine.

5   The charging system output should be 14.5 ± 0.5 volts at 2000 rpm's or more.

6   If the output is as specified, the alternator is functioning properly. If the charging system as a whole is not performing as it should, refer to the appropriate Section and check the voltage regulator/rectifier.

7   Low voltage output may be the result of damaged windings in the alternator stator coils or wiring problems. Make sure all electrical connections are clean and tight, then refer to the appropriate Section and check the alternator stator coil windings and leads for continuity.

29 Alternator stator coil — continuity test

1   If charging system output is low or non-existent, the alternator stator coil windings and leads should be checked for proper continuity. The test can be made with the stator in place on the machine.

2   To gain access to the stator coil wiring harness connector, remove the right side cover, the coolant reservoir and the TCI unit. The reservoir and TCI unit can be moved aside after removing the screw that holds the top of the reservoir in place (you may have to disconnect the upper reservoir hose, but be sure to leave the lower hose in place).

3   Locate the stator coil wiring harness connector (photo) and unplug it (the connector contains three white wires).

4   Using an ohmmeter (preferred) or a continuity test light, check for continuity between each of the three white wires coming from the alternator stator (photo). Continuity should exist between any one wire and each of the others (Yamaha actually specifies a resistance of 0.32 ± 10% ohms).

5   Check for continuity between each of the wires and the engine (photo). *No continuity should exist between any of the wires and the case.*

6   If there is no continuity between any two of the wires, or if there is continuity between the wires and an engine ground, an open circuit or a short exists within the stator coils. Since repair of the stator is not feasible, it must be replaced with a new one.

29.3   Stator coil wiring harness connector location (arrow)

29.4   Checking for continuity between the alternator stator coil winding leads (white wires) (note the automotive-type tab connectors used to hook up the meter leads)

29.5   Checking for the absence of continuity between the alternator stator coil winding leads and the engine (ground)

30.2   Alternator stator mounting screws (arrows)

30 Alternator — removal and installation

1   The alternator components include the stator, which is mounted inside the left engine crankcase cover, and the rotor, which is attached to the left end of the crankshaft.
2   The stator and its wiring harness can be detached from the crankcase cover by removing the three stator and two wire guide mounting screws (photo) (refer to Chapter 5, Section 10 for the procedure to follow when removing the left crankcase cover). When installing the new stator, be sure to use thread-locking compound on the screws and tighten them evenly and securely. Make sure the wiring harness is routed correctly and held in place by the wire guide.
3   Refer to Chapter 2 for the procedure to follow when removing the rotor.

31 Voltage regulator/rectifier — check

1   The voltage regulator/rectifier is attached to the frame directly behind the battery. To gain access to the necessary wiring harness connectors for the check, remove the right side cover, the coolant reservoir and the TCI unit. The reservoir and TCI unit can be moved aside after removing the screw that holds the top of the reservoir in place (you may have to disconnect the upper reservoir hose, but be sure to leave the lower hose in place).
2   Locate the rectifier/regulator wiring harness connectors and unplug them (one connector contains three white wires and is attached to the alternator stator coils; the other connector contains a red, a black, a brown and a yellow wire).
3   To check the rectifier diodes, an ohmmeter is required. Place the ohmmeter selector switch in the Rx10 position. Attach the positive (red) ohmmeter lead to the red wire terminal in the rectifier wiring harness connector. Touch the negative (black) ohmmeter lead, in turn, to each of the white wire terminals in the alternator stator coil wiring harness connector. The ohmmeter must indicate *continuity* at each terminal connection. Reverse the ohmmeter leads (black lead to red wire terminal, red lead to white wire terminals) and repeat the test. The ohmmeter must now indicate *no continuity* at each terminal connection.
4   Repeat the check with the ohmmeter leads attached to the black wire terminal in the rectifier wiring harness connector. With the positive (red) ohmmeter lead attached to the black wire terminal, *no continuity* should exist when the negative (black) ohmmeter lead is connected to the white wire terminals. *Continuity* must be indicated when the black ohmmeter lead is attached to the black wire terminal and the red lead is connected to the white wire terminals.
5   If the results of the check are not exactly as described, one or more of the diodes in the rectifier is faulty. Since the diodes are sealed in the unit, the entire rectifier/regulator assembly must be replaced with a new one.

32 Electric fan — check

1   If the engine overheats as a result of failure of the fan to operate as required, make the following check.
2   Remove the fuel tank and airbox by referring to Chapter 4.
3   Locate the fan wiring harness connector and unplug it (it contains one black and one dark blue wire).
4   Using jumper leads with automotive-type tab connectors (female), connect the fan motor directly to the battery (photo). If the motor does not operate, it is defective and should be replaced with a new one (Chapter 3). If it does operate, check the fan relay, the thermostatic switch, the main fuse and the wiring between the main fuse and the relay (in that order).

33 Electric fan relay and thermostatic switch — check

1   If the engine overheats due to failure of the fan to operate as required, and the fan has been checked and proven to be in good condition, check the relay and thermostatic switch as follows.
2   Locate the relay and disconnect the wiring harness from it (photo). Check for battery voltage at each of the red leads in the wiring harness side of the connector (photo). If no voltage is indicated, check the fan fuse located inside the headlight shell (photo), the main fuse and the wiring between the fuses and between the fan fuse and the relay.
3   If battery voltage is indicated, disconnect the wire (blue/green) from the thermostatic switch and touch it to the T-joint the switch is mounted in (photo). If the fan operates, the relay is in good condition, but the thermostatic switch is defective and must be replaced with a new one (Chapter 3).
4   If the fan does not operate when the wire is grounded on the T-joint, listen carefully for a 'click' at the relay when the wire is grounded. If no noise is heard, replace the relay with a new one, then repeat the checks.

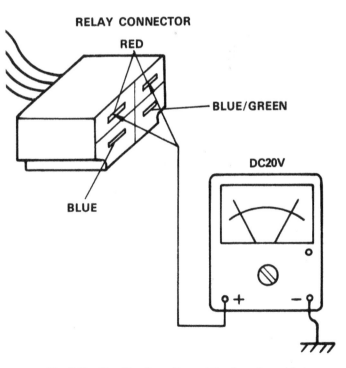

Fig. 8.10 Checking for voltage at the fan relay wiring harness red wire terminals (Sec 33)

32.4 Using female automotive-type tab connectors to apply battery voltage directly to the fan motor wire terminals

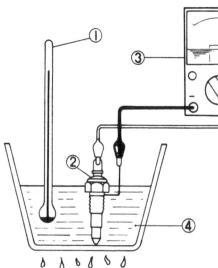

Fig. 8.11 Checking the coolant temperature
sending unit for proper operation (Sec 34)

1   Thermometer
2   Sending unit
3   Ohmmeter
4   Heated water

33.2a   Electric fan relay and connector (arrow) location

33.2b   Checking for battery voltage at the relay harness
connector red wire terminals (the other voltmeter lead is attached
to the negative battery terminal)

33.2c   Fan fuse location

33.3   Checking the thermostatic switch by touching the wire to
the T-joint (the fan should operate)

34 Coolant temperature sending unit — check

1 If the coolant temperature gauge does not operate properly, but the gauge seems to be in good condition (Section 35), refer to Chapter 3 and remove the coolant temperature sending unit.

2 Attach one ohmmeter lead to the sending unit wire terminal and the other lead to the body, then immerse the inner end of the sending unit in a pan of heated water, along with a thermometer (see the accompanying illustration).

Water Temperature	50°C (122°F)	80°C (176°F)	100°C (212°F)
Resistance	153.9Ω	47.5 ~ 56.8Ω	26.2 ~ 29.3Ω

Fig. 8.12 Coolant temperature sending unit resistance at various temperatures (Sec 34)

3 Check the resistance with the water at approximately 122 °F, 176 °F and 212 °F and compare the results to the resistance figures on the accompanying chart. If the readings are not as indicated, replace the sending unit with a new one.

35 Coolant temperature gauge — check

1 To check the gauge meter coil inside the instrument cluster for the correct resistance, the bottom cover of the instrument cluster must be removed (see Section 15).

2 Locate the temperature gauge wires (brown and green/red) and separate them from the meter posts.

3 Place the ohmmeter selector switch in the Rx1 position, then attach the ohmmeter leads to the temperature gauge meter posts and note the resistance reading. Compare the measured resistance to the Specifications. If it is not as specified, the gauge meter is probably faulty and should be replaced with a new one.

Fig. 8.14 Component key for 1982 German models

1 Right-hand handlebar switch
2 Lighting switch
3 Engine stop switch
4 Starter switch
5 Front brake switch
6 Ignition coil
7 Starter motor
8 Starter relay
9 Battery
10 TCI unit
11 Rear brake switch
12 Thermo unit
13 Thermostatic switch
14 Fuse box
15 Starting circuit cut-off relay
16 Electric fan relay
17 Right-hand rear turn signal
18 Tail light
19 License plate light
20 Left-hand rear turn signal
21 Turn signal relay
22 Neutral switch
23 Oil pressure switch
24 Regulator/rectifier
25 Alternator
26 Pickup coils
27 Electric fan motor
28 Ignition coil
29 Frame earth point
30 Clutch switch
31 Headlight dimmer switch
32 Headlight passing switch
33 Horn switch
34 Turn signal switch
35 Left-hand handlebar switch
36 Left-hand front turn signal
37 Highbeam indicator light
38 Oil pressure warning light
39 Neutral indicator light
40 Speed sensor - reed switch
41 Tachometer
42 Engine temperature gauge
43 Left-hand turn signal warning light
44 Right-hand turn signal warning light
45 Instrument light
46 Instrument light
47 Instrument console
48 Ignition main switch
49 Auxiliary light
50 Headlight
51 Horn
52 Right-hand front turn signal

Fig. 8.15 Component key for all models except USA and German

1 Right-hand handlebar switch
2 Lighting switch
3 Engine stop switch
4 Starter switch
5 Front brake switch
6 Ignition coil
7 Starter motor
8 Starter relay
9 Battery
10 TCI unit
11 Rear brake switch
12 Thermo unit
13 Thermostatic switch
14 Fuse box
15 Starting circuit cut-off relay
16 Electric fan relay
17 Right-hand rear turn signal
18 Taillight
19 License plate light
20 Left-hand rear turn signal
21 Turn signal relay
22 Turn signal self-cancelling unit
23 Neutral switch
24 Oil pressure switch
25 Regulator/rectifier
26 Alternator
27 Pickup coils
28 Electric fan motor
29 Ignition coil
30 Frame earth point
31 Clutch switch
32 Headlight dimmer switch
33 Headlight passing switch
34 Horn switch
35 Turn signal switch
36 Left-hand handlebar switch
37 Left-hand front turn signal
38 Highbeam indicator light
39 Oil pressure warning light
40 Neutral indicator light
41 Speed sensor - reed switch
42 Tachometer
43 Engine temperature gauge
44 Left-hand turn signal warning light
45 Right-hand turn signal warning light
46 Instrument light
47 Instrument light
48 Instrument console
49 Ignition main switch
50 Auxiliary light
51 Headlight
52 Horn
53 Right-hand front turn signal

Wiring diagram color key

B	Black	B/Y	Black and yellow
Y	Yellow	R/W	Red and white
L	Blue	Br/W	Brown and white
G	Green	Y/R	Yellow and red
R	Red	L/W	Blue and white
O	Orange	G/Y	Green and yellow
W	White	R/Y	Red and yellow
Br	Brown	L/G	Blue and green
Sb	Light blue	G/R	Green and red
Ch	Chocolate	L/B	Blue and black
Dg	Dark green	Y/G	Yellow and green
B/R	Black and red	Y/B	Yellow and black
B/W	Black and white	W/G	White and green
L/Y	Blue and yellow	Y/L	Yellow and blue

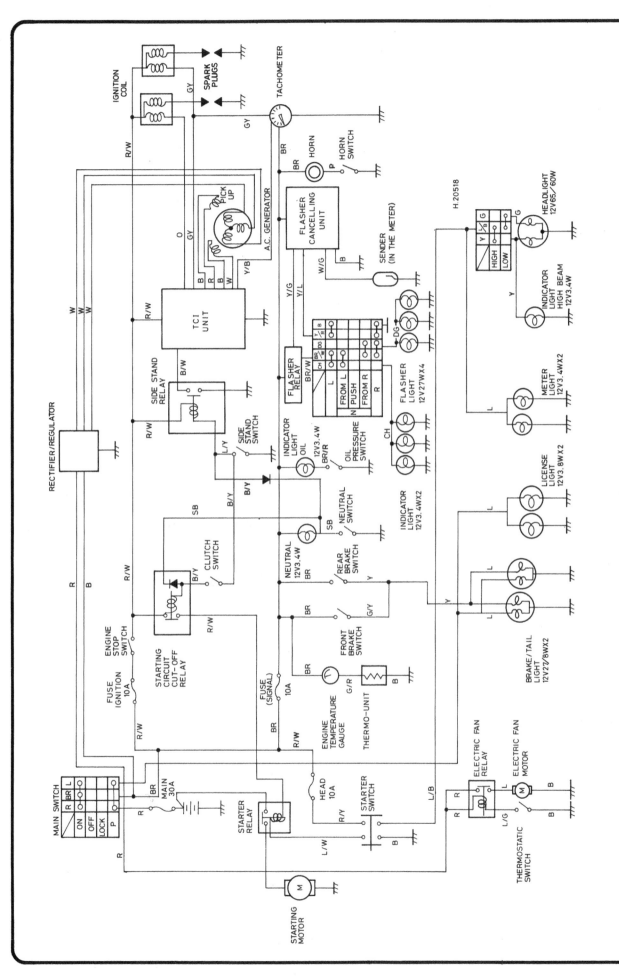

Fig. 8.13 Wiring diagram – 1982 USA models

See page 160 for color key

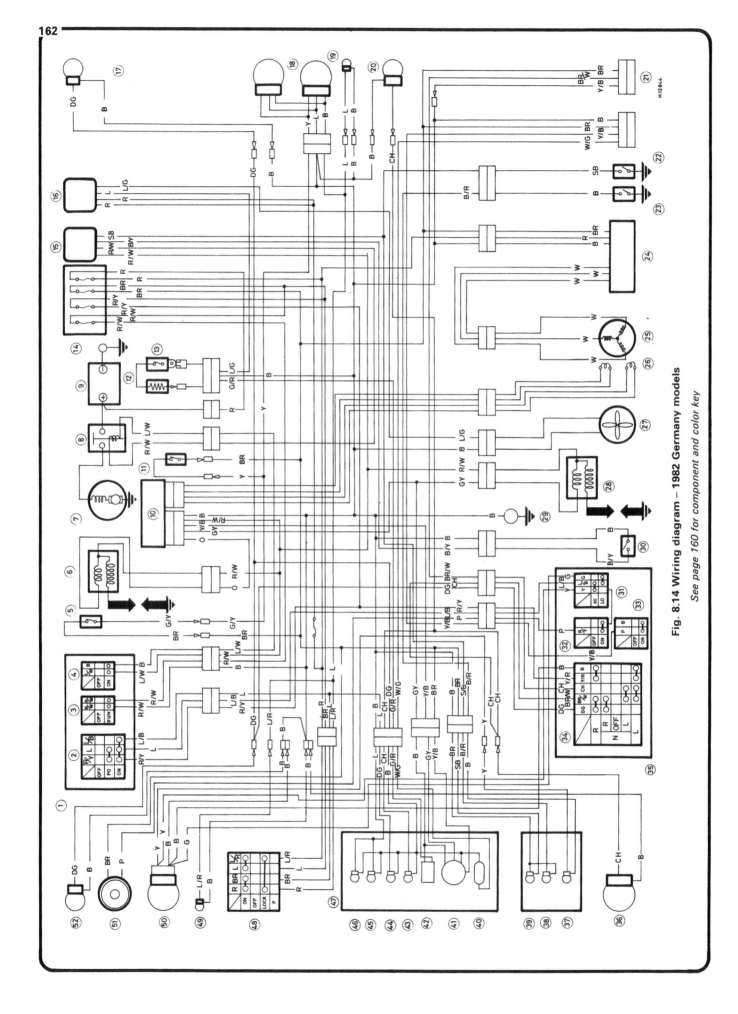

Fig. 8.14 Wiring diagram – 1982 Germany models

See page 160 for component and color key

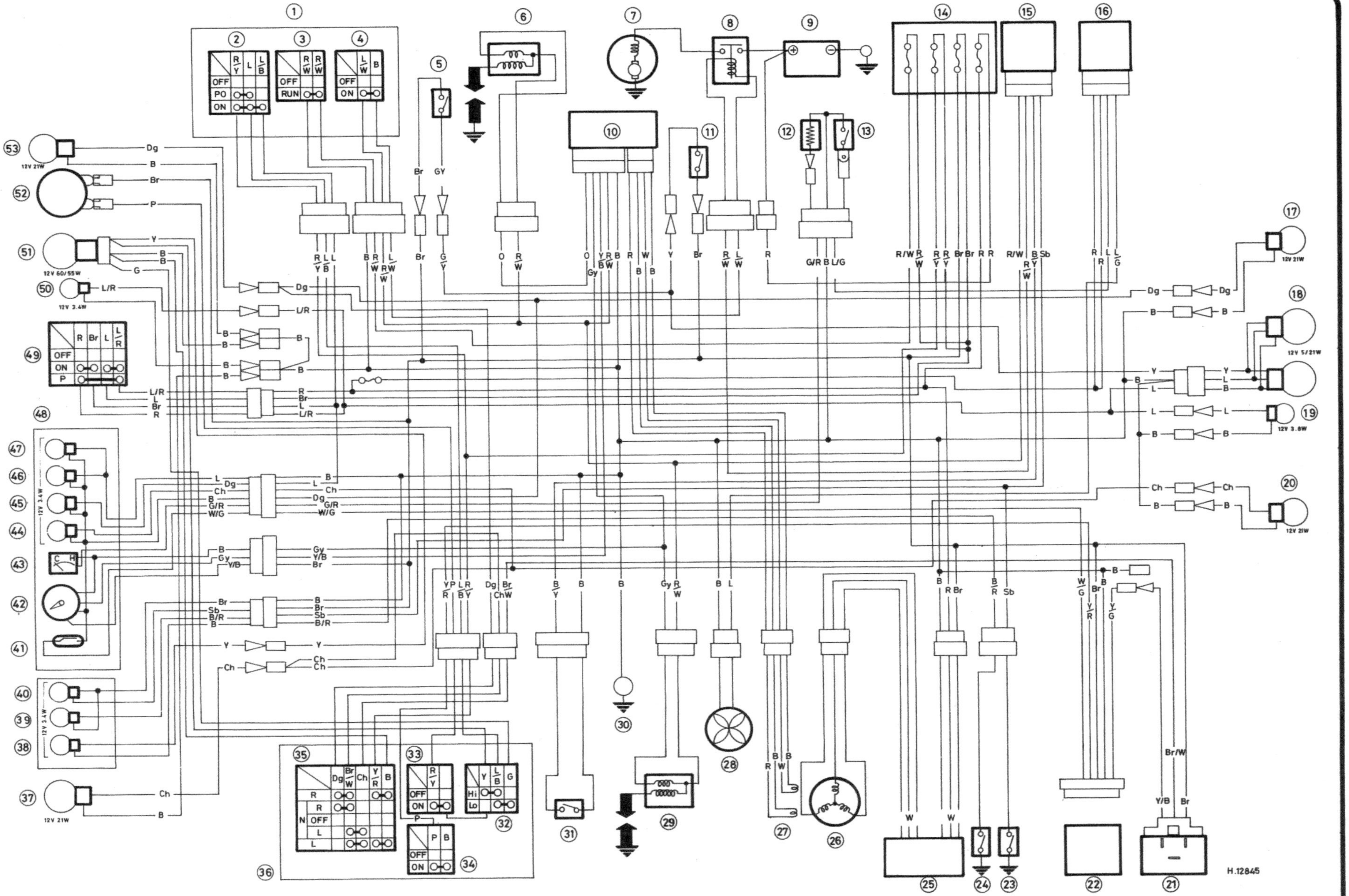

H.12845

Fig. 8.15 Wiring diagram – 1982 models (except USA and Germany)

See page 160 for component and color key

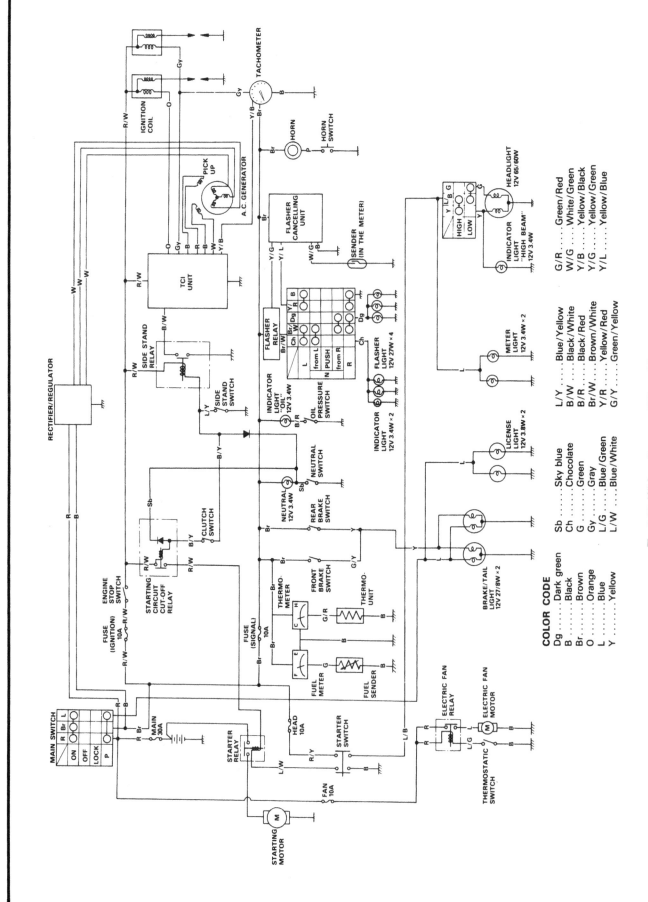

Fig. 8.16 Wiring diagram – RK models

COLOR CODE

Dg	Dark green	Sb	Sky blue	L/Y	Blue/Yellow
B	Black	Ch	Chocolate	B/W	Black/White
Br	Brown	G	Green	B/R	Black/Red
O	Orange	Gy	Gray	Br/W	Brown/White
L	Blue	L/G	Blue/Green	Y/R	Yellow/Red
Y	Yellow	L/W	Blue/White	G/Y	Green/Yellow

G/R	Green/Red
W/G	White/Green
Y/B	Yellow/Black
Y/G	Yellow/Green
Y/L	Yellow/Blue

Chapter 9 Fairing (RK models)

Contents

Specifications

Torque specifications	Ft-lb	Nm
Rear view mirror/upper panel	2.2	3
Signal light	12	17
Lower panel		
Screw	2.2	3
Nut	4.5	6

1 General information

Later (RK) models were equipped with a frame-mounted three-quarter fairing. The front turn signal lights and the rear view mirrors are mounted on the upper fairing panel.

Several maintenance and repair operations require removal of part or all of the fairing – be sure to read Section 2 for additional information.

2 Procedures requiring fairing removal

The fairing assembly must be removed to perform certain maintenance and major repair procedures. Some procedures require only part of the fairing to be removed. However, Yamaha recommends removal of the entire fairing whenever service is performed. This will ensure the fairing is not scratched or otherwise damaged. The accompanying table (Fig. 9.1) lists the fairing components that must be removed to gain access to a particular area or component.

	Section to be removed	Side cover (LH)	Side cover (RH)	Seat (open)	Fuel tank	Fairing Assembly: Upper panel assembly	Fairing Assembly: Lower panel (LH)	Fairing Assembly: Lower panel (RH)
ENGINE	Spark plug replacement						O	
	Valve clearance adjustment (front)	O	O	O	O		O	O
	Valve clearance adjustment (rear)	O	O	O	O			
	Cooling system (coolant replacement)						O	O
	Air cleaner service	O	O	O	O			
	Engine oil service							
	Compression pressure check						O	
	Carburetor synchronization							
	Engine removal	O	O	O	O		O	O
CHASSIS	Front fork oil replacement							
	Rear shock absorber (spring adjustment)	O	O	O	O			
	Rear shock absorber (damper adjustment)		O					
ELEC.	Battery service	O						
	Headlight (bulb replacement)							
	Ignition timing check							

Fig. 9.1 Remove the fairing component(s) marked with a circle to perform the procedures listed in the left-hand column

3   Fairing – removal and installation

1   Remove the nut/bolt connecting both lower panels at the bottom (see Fig. 9.2).
2   Remove the three remaining mounting screws from each panel and detach both lower panels.
3   Unplug the electrical connectors from both front turn signals.
4   Remove the rear-view mirrors from the upper panel.
5   Pull the upper panel assembly toward the front and remove it.
6   Installation is the reverse of removal.

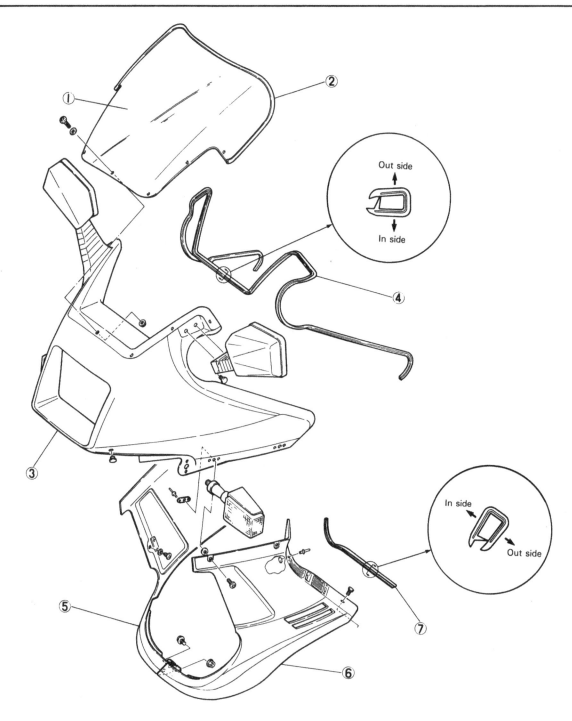

Fig. 9.2 Fairing components – exploded view

1	Windscreen	5	Lower panel (right)
2	Cowling trim	6	Lower panel (left)
3	Upper panel	7	Molding 2
4	Molding 1		

Conversion factors

Length (distance)

Inches (in)	X	25.4	= Millimetres (mm)	X 0.0394	= Inches (in)
Feet (ft)	X	0.305	= Metres (m)	X 3.281	= Feet (ft)
Miles	X	1.609	= Kilometres (km)	X 0.621	= Miles

Volume (capacity)

Cubic inches (cu in; in³)	X	16.387	= Cubic centimetres (cc; cm³)	X 0.061	= Cubic inches (cu in; in³)
Imperial pints (Imp pt)	X	0.568	= Litres (l)	X 1.76	= Imperial pints (Imp pt)
Imperial quarts (Imp qt)	X	1.137	= Litres (l)	X 0.88	= Imperial quarts (Imp qt)
Imperial quarts (Imp qt)	X	1.201	= US quarts (US qt)	X 0.833	= Imperial quarts (Imp qt)
US quarts (US qt)	X	0.946	= Litres (l)	X 1.057	= US quarts (US qt)
Imperial gallons (Imp gal)	X	4.546	= Litres (l)	X 0.22	= Imperial gallons (Imp gal)
Imperial gallons (Imp gal)	X	1.201	= US gallons (US gal)	X 0.833	= Imperial gallons (Imp gal)
US gallons (US gal)	X	3.785	= Litres (l)	X 0.264	= US gallons (US gal)

Mass (weight)

Ounces (oz)	X	28.35	= Grams (g)	X 0.035	= Ounces (oz)
Pounds (lb)	X	0.454	= Kilograms (kg)	X 2.205	= Pounds (lb)

Force

Ounces-force (ozf; oz)	X	0.278	= Newtons (N)	X 3.6	= Ounces-force (ozf; oz)
Pounds-force (lbf; lb)	X	4.448	= Newtons (N)	X 0.225	= Pounds-force (lbf; lb)
Newtons (N)	X	0.1	= Kilograms-force (kgf; kg)	X 9.81	= Newtons (N)

Pressure

Pounds-force per square inch (psi; lbf/in²; lb/in²)	X	0.070	= Kilograms-force per square centimetre (kgf/cm²; kg/cm²)	X 14.223	= Pounds-force per square inch (psi; lbf/in²; lb/in²)
Pounds-force per square inch (psi; lbf/in²; lb/in²)	X	0.068	= Atmospheres (atm)	X 14.696	= Pounds-force per square inch (psi; lbf/in²; lb/in²)
Pounds-force per square inch (psi; lbf/in²; lb/in²)	X	0.069	= Bars	X 14.5	= Pounds-force per square inch (psi; lbf/in²; lb/in²)
Pounds-force per square inch (psi; lbf/in²; lb/in²)	X	6.895	= Kilopascals (kPa)	X 0.145	= Pounds-force per square inch (psi; lbf/in²; lb/in²)
Kilopascals (kPa)	X	0.01	= Kilograms-force per square centimetre (kgf/cm²; kg/cm²)	X 98.1	= Kilopascals (kPa)
Millibar (mbar)	X	100	= Pascals (Pa)	X 0.01	= Millibar (mbar)
Millibar (mbar)	X	0.0145	= Pounds-force per square inch (psi; lbf/in²; lb/in²)	X 68.947	= Millibar (mbar)
Millibar (mbar)	X	0.75	= Millimetres of mercury (mmHg)	X 1.333	= Millibar (mbar)
Millibar (mbar)	X	0.401	= Inches of water (inH₂O)	X 2.491	= Millibar (mbar)
Millimetres of mercury (mmHg)	X	0.535	= Inches of water (inH₂O)	X 1.868	= Millimetres of mercury (mmHg)
Inches of water (inH₂O)	X	0.036	= Pounds-force per square inch (psi; lbf/in²; lb/in²)	X 27.68	= Inches of water (inH₂O)

Torque (moment of force)

Pounds-force inches (lbf in; lb in)	X	1.152	= Kilograms-force centimetre (kgf cm; kg cm)	X 0.868	= Pounds-force inches (lbf in; lb in)
Pounds-force inches (lbf in; lb in)	X	0.113	= Newton metres (Nm)	X 8.85	= Pounds-force inches (lbf in; lb in)
Pounds-force inches (lbf in; lb in)	X	0.083	= Pounds-force feet (lbf ft; lb ft)	X 12	= Pounds-force inches (lbf in; lb in)
Pounds-force feet (lbf ft; lb ft)	X	0.138	= Kilograms-force metres (kgf m; kg m)	X 7.233	= Pounds-force feet (lbf ft; lb ft)
Pounds-force feet (lbf ft; lb ft)	X	1.356	= Newton metres (Nm)	X 0.738	= Pounds-force feet (lbf ft; lb ft)
Newton metres (Nm)	X	0.102	= Kilograms-force metres (kgf m; kg m)	X 9.804	= Newton metres (Nm)

Power

Horsepower (hp)	X	745.7	= Watts (W)	X 0.0013	= Horsepower (hp)

Velocity (speed)

Miles per hour (miles/hr; mph)	X	1.609	= Kilometres per hour (km/hr; kph)	X 0.621	= Miles per hour (miles/hr; mph)

Fuel consumption*

Miles per gallon, Imperial (mpg)	X	0.354	= Kilometres per litre (km/l)	X 2.825	= Miles per gallon, Imperial (mpg)
Miles per gallon, US (mpg)	X	0.425	= Kilometres per litre (km/l)	X 2.352	= Miles per gallon, US (mpg)

Temperature

Degrees Fahrenheit = (°C x 1.8) + 32   Degrees Celsius (Degrees Centigrade; °C) = (°F - 32) x 0.56

*It is common practice to convert from miles per gallon (mpg) to litres/100 kilometres (l/100km),
where mpg (Imperial) x l/100 km = 282 and mpg (US) x l/100 km = 235

Index